D1265017

Statistical Properties of Scattered Light

QUANTUM ELECTRONICS — PRINCIPLES AND APPLICATIONS

A Series of Monographs

EDITED BY
YOH-HAN PAO
Case Western Reserve University
Cleveland, Ohio

Statistical Properties of Scattered Light

Bruno Crosignani and Paolo Di Porto

Laser Laboratory, Fondazione Ugo Bordoni
Istituto Superiore Poste e Telecomunicazioni
Rome, Italy

Mario Bertolotti

Istituto di Fisica della Facoltà d'Ingeneria
Università di Roma
Rome, Italy

ACADEMIC PRESS   New York   San Francisco   London   1975

A Subsidiary of Harcourt Brace Jovanovich, Publishers

ACADEMIC PRESS, INC.
111 Fifth Avenue, New York, New York 10003

United Kingdom Edition published by
ACADEMIC PRESS, INC. (LONDON) LTD.
24/28 Oval Road, London NW1

Library of Congress Cataloging in Publication Data

Crosignani, Bruno.
    Statistical properties of scattered light.

    (Quantum electronics series)
    Bibliography: p.
    Includes indexes.
    1.   Light—Scattering.   2.   Statistical mechanics.
3.   Photons—Measurement.   4.   Light beating spectroscopy.
I.   Di Porto, Paolo, joint author.   II.   Bertolotti,
Mario, joint author.   III.   Title.
QC427.4.C76   535   74-27778
ISBN 0–12–199050–8

Contents

Chapter III.   **Macroscopic Approach to Scattering Theory**

Chapter IV.   **Statistical Description of Electromagnetic Radiation**

Chapter V.   **Statistical Properties of Light Scattered by Fluids and Plasmas**

Chapter VI.   **Statistical Properties of Light Scattered by Small Particles**

Preface

The object of this book is to call to the attention of researchers engaged in light-scattering experiments the potential that the new techniques of photon counting and light-beating spectroscopy offer. To do so, we present the conditions and circumstances whereby this new class of experiments is able to furnish information not contained, even in principle, in ordinary measurements. We do not compare the experimental advantages of these techniques with the traditional ones, but rather we bring together and treat in a unified way the relevant situations in which the new methods prove particularly valid.

As the basic nature of the scattering process in important for these discussions, and as this subject is far from being thoroughly investigated, we felt it useful and necessary to devote some of this work to a summary of the existing theories and to try to clarify some aspects of the specific problems such as the connection between macroscopic and microscopic approaches.

The book offers a concise, unified treatment of various kinds of scattering (from liquids, from plasmas, from macroscopic particles), which we hope will make it useful not only to physicists and electrical engineers who are active in the field but to researchers who wish to familiarize themselves with this topic. In this respect, the authors have tried to render the book as self-sufficient as possible by devoting some space to the quantization of the electromagnetic field and to the introduction of some basic concepts of statistical mechanics.

The authors wish to thank Professor R.J. Glauber for his encouragement to write this book. They are grateful to Professor S.H. Chen for discussing with them the plan of the book and to Professor E. Wolf for a useful suggestion concerning the macroscopic dielectric constant.

Finally, they desire to acknowledge the patience and ability with which Miss A. De Cresce has typed the manuscript, the help of Mr. C. Sanipoli in preparing the original illustrations, and Carlo Crosignani for his special assistance.

I

The Electromagnetic Field Scattered by a System of Charges

I.1 Historical Introduction

The first scientific light scattering experiment was performed by Tyndall (1869a, b), who observed the natural light scattered by particles whose size was small compared with the incident wavelength and noticed the appearance of a bluish hue in the scattered radiation. Rayleigh (1881) gave the theoretical explanation of this fact, showing that the intensity of light scattered by such particles, considered noninteracting, is inversely proportional to the fourth power of the wavelength. This effect accounts in particular for the blue color of the sky, once the scattering centers are assumed to be the molecules constituting the atmosphere, possessing a mean free path larger than the optical wavelength.

The acknowledgment of the existence of the scattering process

1

at the molecular level opened the way to the investigation of light scattering from dense media such as liquids, in which the constituting molecules cannot be considered as independent and the mean free path does not exceed the wavelength. In this case, the intensities of the microfields scattered by the single molecules no longer combine, but one can take advantage of the presence of many molecules in a volume whose linear dimensions are of the order of the wavelength, in order to develop a theory in which the scattering mechanism is ascribed to dielectric constant fluctuations due to thermal agitation (Smolouchosky, 1908; Einstein, 1910).

The next step is to recognize the influence of the temporal behavior of the thermal fluctuations in changing the frequency of the scattered light. From a quantitative point of view, the most relevant change [apart from "combination scattering" (Raman, 1928) dealing with frequency shifts related to the internal dynamics of molecules] is associated with two symmetrical frequency shifts around the unperturbed frequency (Mandel'shtam–Brillouin doublet), which are connected with the propagation of sound waves (pressure fluctuations) in the scattering medium, an effect first predicted by Brillouin (1922) and Mandel'shtam (1926) and experimentally observed by Gross (1930). The damping of the nonpropagating entropy fluctuations as well as the finite lifetime of the sound waves are, respectively, responsible for the presence of a broadening around the central frequency (Rayleigh line) and of the Mandel'shtam–Brillouin doublet, whose structure can be investigated on the basis of the theoretical method furnished by Landau and Placzek (1934). Accurate measurements of the profile of the Rayleigh and Brillouin lines were at that time beyond the resolving power of existing instruments and, in fact, only became feasible (Lastovka and Benedek, 1966; Greytak and Benedek, 1966) after the introduction of laser sources.

From a theoretical point of view, more sophisticated approaches to the investigation of the spectrum of the scattered light were provided by Rytov (1957) and by Komarov and Fisher (1962), respectively, on macroscopic and microscopic bases, the last method being a direct generalization of the formalism introduced by Glauber (1952, 1954) and van Hove (1954) for neutron scattering.

From an experimental point of view, a dramatic advance was due to the development of a new powerful spectroscopic technique known as "light-beating spectroscopy" (Forrester *et al.*, 1955). This experimental method, employed in both the hetero-dyne and self-beating approaches, permits the measurement of small frequency shifts of optical radiation scattered by macro-scopic particles (Cummins *et al.*, 1964) and large molecules in solution (Ford and Benedek, 1965).

While the conventional methods for investigating scattered light are based on intensity measurements, new possibilities have arisen in connection with the more general statistical description of electromagnetic radiation in terms of higher-order correlation functions (Wolf, 1963; Glauber, 1963a, b). In this framework, a detailed analysis of the scattered light can be achieved by means of photon-counting experiments (Freed and Haus, 1966; Johnson *et al.*, 1966; Arecchi, 1965) or more sophisticated digital techniques (Pike and Jakeman, 1974). They greatly enhance the information that can be gained on the scattering medium, as will become apparent in the course of this book.

The reader is referred to the book by Fabelinskii (1968) for an exhaustive treatment of conventional light scattering, while the methods dealing with modern techniques are extensively treated in the Proceedings of the NATO Advanced Study Institute held in Capri, Italy, 1973 (Cummins and Pike, 1974). Other recent valuable sources of information are furnished by the review papers by Fleury and Boon (1973) and Gelbart (1974), respec-tively, on scattering by fluids and depolarized light scattering by simple fluids, and the book by Chu (1974), which mainly deals with quasielastic scattering.

I.2 The Electromagnetic Field Generated by a Distribution of Charges

The interaction of an external electromagnetic radiation with a material system gives rise to a scattered field, which is composed of microfields scattered by the elementary electric charges present in the medium. Therefore, the study of the observed field requires a preliminary investigation of the contribution pertinent to the single charges. For a large class of systems, the charges are distributed among many molecules, which are subsystems maintaining their individuality in time. Whenever the wavelength of the incident radiation is large compared to the characteristic dimension of a molecule, it reacts as a whole to the electromagnetic perturbation, thus becoming the effective scattering center. One has then to investigate the behavior of the single molecule under the influence of an electromagnetic radiation (which will also include the field generated by the other molecules) in order to evaluate the total scattered field. In the following, we shall discuss the deterministic electromagnetic field scattered by a system of molecules.

The treatment of the electromagnetic radiation interacting with a material system hinges on Maxwell's equations in vacuo, in the presence of the appropriate distribution of charges and currents. This fundamental, or microscopic, approach is the basis for justifying all treatments in which a less detailed description of the electromagnetic behavior of the medium is used.

Although the microscopic approach is exact in principle, a number of approximations often has to be introduced in order to simplify the analytical procedure. This holds true for most scattering phenomena, whose complexity depends on the characteristics of the scattering medium, as well as on the frequency and intensity of the incident radiation. In order to examine the main features of the microscopic treatment, we shall first summarize some classical notions about the field generated by a given system of charges and currents, which will then be identified with those pertaining to the molecules of the scattering medium.

In Gaussian units, which will be used throughout this book, Maxwell's equations in vacuo are (Stratton, 1941, Chapter 1)

$$\nabla \times \mathbf{E} = -\frac{1}{c}\frac{\partial \mathbf{H}}{\partial t} \tag{1-1}$$

$$\nabla \times \mathbf{H} = \frac{4\pi}{c}\mathbf{j} + \frac{1}{c}\frac{\partial \mathbf{E}}{\partial t} \tag{1-2}$$

$$\nabla \cdot \mathbf{E} = 4\pi\rho \tag{1-3}$$

$$\nabla \cdot \mathbf{H} = 0 \tag{1-4}$$

where $\mathbf{E}$ and $\mathbf{H}$ are, respectively, the electric and magnetic vectors, and c represents the velocity of light in vacuo. As a consequence of the two nonhomogeneous Eqs. (1-2) and (1-3), the charge ρ and current density $\mathbf{j}$ obey the equation of continuity

$$\frac{\partial \rho}{\partial t} + \nabla \cdot \mathbf{j} = 0 \tag{1-5}$$

If one expresses ρ and $\mathbf{j}$ in terms of a single vector $\mathbf{P}$,

$$\rho = -\nabla \cdot \mathbf{P}, \qquad \mathbf{j} = \partial \mathbf{P}/\partial t \tag{1-6}$$

the field equations take the form

$$\nabla \times \mathbf{E} = -\frac{1}{c}\frac{\partial \mathbf{H}}{\partial t} \tag{1-7}$$

$$\nabla \times \mathbf{H} = \frac{4\pi}{c}\frac{\partial \mathbf{P}}{\partial t} + \frac{1}{c}\frac{\partial \mathbf{E}}{\partial t} \tag{1-8}$$

$$\nabla \cdot \mathbf{E} = -4\pi\nabla \cdot \mathbf{P} \tag{1-9}$$

$$\nabla \cdot \mathbf{H} = 0 \tag{1-10}$$

while the equation of continuity is identically satisfied.

The solutions of these equations can be written as

$$E = \nabla(\nabla \cdot \Pi) - \frac{1}{c^2} \frac{\partial^2 \Pi}{\partial t^2} \qquad (1\text{-}11)$$

$$H = \frac{1}{c} \nabla \times \frac{\partial \Pi}{\partial t} \qquad (1\text{-}12)$$

where Π, which is known as the *Hertz vector*, must satisfy the equation

$$\nabla^2 \Pi - \frac{1}{c^2} \frac{\partial^2 \Pi}{\partial t^2} = -4\pi P \qquad (1\text{-}13)$$

The general solution of Eq. (1-13) is

$$\Pi(r, t) = \int_V \frac{P[r', t-(R/c)]}{R} dr' + \Pi_0(r, t) \qquad (1\text{-}14)$$

where r is the observation point, $R = |r - r'|$, V represents the volume occupied by the sources, and Π_0 obeys the homogeneous equation associated with Eq. (1-13) and represents the electromagnetic field that would be present in the absence of sources. Since this field is not the object of our interest, we shall not consider Π_0 here.

Let us consider the Fourier transform in time of $P(r', t)$:

$$P(r', t) = \int_{-\infty}^{+\infty} P_\omega(r') e^{-i\omega t} d\omega \qquad (1\text{-}15)$$

and let us assume that for all significant frequencies the condition

$$\frac{\omega l}{c} = \frac{2\pi l}{\lambda} \ll 1 \qquad (1\text{-}16)$$

is satisfied, l representing a characteristic linear dimension of the volume occupied by the charges. Under this hypothesis, a suitable development in a series of *multipoles* can be given for the Fourier transform of $\Pi(\mathbf{r}, t)$:

$$\Pi_\omega(\mathbf{r}) = \int_V \mathbf{P}_\omega(\mathbf{r}') \frac{e^{ikR}}{R} \, d\mathbf{r}' \qquad (1\text{-}17)$$

where $k = \omega/c$ (Stratton, 1941, Chapter VIII; Phillips, 1962). Let us consider here only the first two terms $\Pi_\omega^{(0)}$ and $\Pi_\omega^{(1)}$ of the field expansion. One has

$$\Pi_\omega^{(0)}(\mathbf{r}) = \mathbf{P}_\omega^{(1)} \frac{e^{ikr}}{r} \qquad (1\text{-}18)$$

with

$$\mathbf{P}_\omega^{(1)} = \int_V \mathbf{r}' \rho_\omega \, d\mathbf{r}' \qquad (1\text{-}19)$$

and

$$\Pi_\omega^{(1)}(\mathbf{r}) = \left(\mathbf{M}_\omega^{(1)} \times \mathbf{\eta} - \frac{i\omega}{2c} \mathbf{P}_\omega^{(2)} \cdot \mathbf{\eta} \right) \left(\frac{1}{r} + \frac{i}{kr^2} \right) e^{ikr} \qquad (1\text{-}20)$$

where $\mathbf{\eta} = \mathbf{r}/r$ and

$$\mathbf{M}_\omega^{(1)} = \frac{1}{2c} \int_V \mathbf{r}' \times \mathbf{j}_\omega(\mathbf{r}') \, d\mathbf{r}' \qquad (1\text{-}21)$$

$$\mathbf{P}_\omega^{(2)} = \int_V \mathbf{r}' \mathbf{r}' \rho_\omega(\mathbf{r}') \, d\mathbf{r}' \qquad (1\text{-}22)$$

with ρ_ω and $\mathbf{j}_\omega$ the Fourier time transforms of ρ and $\mathbf{j}$. We observe that $\mathbf{P}_\omega^{(1)}$, $\mathbf{M}_\omega^{(1)}$, and $\mathbf{P}_\omega^{(2)}$ are, respectively, the Fourier transforms of the *electric dipole moment vector* $\mathbf{P}^{(1)}$, *of the magnetic dipole moment vector* $\mathbf{M}^{(1)}$, and of the *electric quadrupole moment tensor* $\mathbf{P}^{(2)}$.[†]

If we introduce the microscopic charge and current densities, defined by

$$\rho(\mathbf{r}, t) = \sum_i q_i \, \delta\left[\mathbf{r} - \mathbf{R}_{1i}(t) \right] \qquad (1\text{-}23)$$

$$\mathbf{j}(\mathbf{r}, t) = \sum_i q_i \, \mathbf{v}_i(t) \, \delta\left[\mathbf{r} - \mathbf{R}_{1i}(t) \right] \qquad (1\text{-}24)$$

[†] From now on, the symbol $\mathbf{AB}$ ($\mathbf{A}$ and $\mathbf{B}$ being two vectors) indicates the tensor $\mathbf{T}$, whose components are $T_{ij} = A_i B_j$, while the symbol $\mathbf{T} \cdot \mathbf{A}$ indicates the vector $\mathbf{C}$, whose components are $C_i = \sum_{k=1}^3 T_{ik} A_k$.

where $\mathbf{R}_{1i}(t)$ and $\mathbf{v}_i(t)$ are, respectively, the position and the velocity of the ith particle, q_i its charge, and the symbol δ stands for the usual Dirac delta function, we then have

$$\mathbf{P}^{(1)}(t) = \sum_i q_i \mathbf{R}_{1i}(t) \tag{1-25}$$

$$\mathbf{M}^{(1)}(t) = \frac{1}{2c} \sum_i q_i \mathbf{R}_{1i}(t) \times \mathbf{v}_i(t) \tag{1-26}$$

$$\mathbf{P}^{(2)}(t) = \sum_i q_i \mathbf{R}_{1i}(t) \mathbf{R}_{1i}(t) \tag{1-27}$$

The system of charges with which we shall actually deal is a molecule that consists of point charges whose motion determines its electromagnetic properties, that is, its electric and magnetic multipoles. In particular, the dipole moment that is present in the absence of any external perturbation is referred to as *permanent*. It is, for example, zero whenever the structure of the molecule is such that the center of mass of the electrons is coincident with that of the positive charges (*nonpolar* molecules). We are interested in the behavior of the molecule when it is acted on by an external radiation field. More precisely, our aim is to evaluate the field generated by the molecule in the surrounding space in the presence of the external perturbation. This field is not influenced by the presence of permanent multipoles if the frequency of the incident perturbation is large enough (as actually happens in the optical range), since this would imply a rotation of the whole molecule, which is too heavy to follow the rapidly varying electric field. Thus the scattered field is associated with the multipoles induced by the external field, so that $\mathbf{P}^{(1)}$, $\mathbf{M}^{(1)}$, $\mathbf{P}^{(2)}$, etc., must be thought of as deviations from their permanent values.

In order to give an estimate of the relative importance of the various multipolar terms with respect to the dipole term, let us assume as a first approximation that the external perturbation is seen as monochromatic by the molecule. This would be the case for a molecule at rest in the presence of a monochromatic perturbation; in an actual scattering experiment the molecule is moving and receives, beyond a monochromatic field coming from outside the material system, the microfields scattered by the other moving molecules. In any case, for a nonrelativistic medium, that is, for molecular velocities much less than the velocity of light, the relative frequency spread is small enough that the monochromatic approximation can be reasonably assumed for use in estimating the multipole terms.

In a linear approximation scheme (which we shall adopt in the following) the deviation of **P** is proportional to the external perturbation, so that all the multipoles possess a harmonic time behavior with the field frequency ω_0. This allows us to invert immediately Eqs. (1-18) and (1-20) to give

$$\boldsymbol{\Pi}^{(0)}(\mathbf{r}, t) = \mathbf{P}_0^{(1)} \frac{\exp(ik_0 r - i\omega_0 t)}{r} \tag{1-28}$$

with

$$\mathbf{P}_0^{(1)} e^{-i\omega_0 t} = \sum_i q_i \mathbf{R}'_{1i}(t) \tag{1-29}$$

and

$$\boldsymbol{\Pi}^{(1)}(\mathbf{r}, t) = \left(\mathbf{M}_0^{(1)} \times \boldsymbol{\eta} - \frac{i\omega_0}{2c} \mathbf{P}_0^{(2)} \cdot \boldsymbol{\eta} \right) \left(\frac{1}{r} + \frac{i}{k_0 r^2} \right)$$
$$\times \exp(ik_0 r - i\omega_0 t) \tag{1-30}$$

with

$$\mathbf{M}_0^{(1)} e^{-i\omega_0 t} = \frac{1}{2c} \sum_i q_i [\mathbf{R}'_{1i}(t) + \mathbf{R}_{0i}] \times \mathbf{v}'_i(t)$$
$$\simeq \frac{1}{2c} \sum_i q_i \mathbf{R}_{0i} \times \mathbf{v}'_i(t) \tag{1-31}$$

and

$$\mathbf{P}_0^{(2)} e^{-i\omega_0 t} = \sum_i \{ q_i [\mathbf{R}'_{1i}(t) + \mathbf{R}_{0i}][\mathbf{R}'_{1i}(t) + \mathbf{R}_{0i}] - q_i \mathbf{R}_{0i} \mathbf{R}_{0i} \}$$
$$\simeq \sum_i q_i [\mathbf{R}'_{1i}(t) \mathbf{R}_{0i} + \mathbf{R}_{0i} \mathbf{R}'_{1i}(t)] \tag{1-32}$$

The left-hand sides of Eqs. (1-29), (1-31), and (1-32) define the induced electric dipole, magnetic dipole, and electric quadrupole moments, respectively, under the influence of an external harmonic perturbation of frequency ω_0, which gives rise to a harmonic velocity $\mathbf{v}'_i(t)$ corresponding to a deviation $\mathbf{R}'_{1i}(t)$ from the unperturbed value $\mathbf{R}_{0i}$ of the position of the ith charge. We observe that the terms $\mathbf{R}'_{1i}(t) \times \mathbf{v}'_i(t)$, $\mathbf{R}'_{1i}(t)\,\mathbf{R}'_{1i}(t)$ have been neglected in Eqs. (1-31) and (1-32) since they are quadratic in the perturbation field. Furthermore we have considered the unperturbed positions $\mathbf{R}_{0i}$ to be independent of time, which entails the harmonic behavior of Eq. (1-30).

In order to evaluate the relative importance of the term $\mathbf{\Pi}^{(1)}$ compared with $\mathbf{\Pi}^{(0)}$, we resort to a slight generalization of the simple classical model adopted for evaluating the induced dipole moment for nonpolar molecules (Born and Wolf, 1970, Chapter II). The molecule is described as an electron bound by an elastic force

$$\mathbf{F} = -s(\mathbf{r} - \mathbf{r}_0) \tag{1-33}$$

to a distance $\mathbf{r}_0$ from the center of positive charges. Denoting by e the electron charge, the permanent dipole moment is given by $e\mathbf{r}_0$, so that $\mathbf{r}_0 = 0$ for nonpolar molecules. Let us consider a situation in which a monochromatic electric field $\mathbf{E}_0 \exp(-i\omega_0 t)$ impinges on the molecule, thus causing the electron to move from its equilibrium position.

The steady-state solution of the equation of motion

$$m\frac{d^2\mathbf{r}}{dt^2} + \gamma\frac{d\mathbf{r}}{dt} + s(\mathbf{r} - \mathbf{r}_0) = e\mathbf{E}_0\,e^{-i\omega_0 t} \tag{1-34}$$

where the term $\gamma\,d\mathbf{r}/dt$ is introduced to take into account the *self-force* of the electron deriving from the conversion of its mechanical energy into radiated energy, is

$$\mathbf{r}(t) = \mathbf{r}_0 + \frac{(e/m)\,\mathbf{E}_0\,e^{-i\omega_0 t}}{(\omega_s^2 - \omega_0^2) + i\omega_0\gamma} \tag{1-35}$$

with $\omega_s = (s/m)^{1/2}$. It is worth noting that the general solution of Eq. (1-34) is obtained by adding to Eq. (1-35) the general solution of the homogeneous equation associated with Eq. (1-34), which can be neglected since it has a transient behavior and vanishes for times larger than $1/\gamma$. Substituting Eq. (1-35) into Eqs. (1-29), (1-31), and (1-32), where the sum is now limited to one electron with trajectory $\mathbf{r}(t)$, we obtain

$$\mathbf{P}_0^{(1)} = \frac{e^2}{m} \frac{1}{(\omega_s^2 - \omega_0^2) + i\omega_0\gamma} \mathbf{E}_0 \qquad (1\text{-}36)$$

$$\mathbf{M}_0^{(1)} = \frac{i\omega_0}{2c} \mathbf{P}_0^{(1)} \times \mathbf{r}_0 \qquad (1\text{-}37)$$

$$\mathbf{P}_0^{(2)} = \mathbf{P}_0^{(1)}\mathbf{r}_0 + \mathbf{r}_0\mathbf{P}_0^{(1)} \qquad (1\text{-}38)$$

After substituting Eqs. (1-37) and (1-38) into (1-30), a comparison with Eq. (1-28) shows that the relative magnitude of the term of $\mathbf{\Pi}^{(1)}$ proportional to $1/r$ with respect to $\mathbf{\Pi}^{(0)}$ is of the order of r_0/λ_0, which is small since we have assumed the wavelength $\lambda_0 = 2\pi c/\omega_0$ of the incident radiation much larger than the linear dimension l of the molecules, which in turn satisfies the relation

$$r_0 \ll l \qquad (1\text{-}39)$$

We note that the relation expressed by Eq. (1-39), which is obviously satisfied for nonpolar molecules ($\mathbf{r}_0 = 0$), holds practically true for most polar molecules, since r_0 represents in practice the distance between the centers of positive and negative charges. The ratio between the term of $\mathbf{\Pi}^{(1)}$ proportional to $1/r^2$ and $\mathbf{\Pi}^{(0)}$ is of the order of r_0/r, which obviously cannot exceed the value r_0/l. Therefore, one can retain, when evaluating the field generated by a molecule in the presence of an external harmonic perturbation, only the contribution due to the induced dipole moment. In effect, the procedure used can be easily generalized in order to justify the omission of all order multipole contributions, associated with the $\mathbf{\Pi}^{(n)}$'s with $n > 1$.

I.3 Molecular Dipole Moment

According to the conclusions of the preceding section, only the knowledge of the dipole moment of the molecule is needed to determine the field radiated under the influence of an external monochromatic radiation field of frequency ω_0. This quantity has been shown, by means of a schematic classical model, to obey the relation [see Eq. (1-36); hereafter we shall write $\mathbf{P}$ instead of $\mathbf{P}^{(1)}$]

$$\mathbf{P}(t) = \alpha(\omega_0)\,\mathbf{E}_0\,e^{-i\omega_0 t} \tag{1-40}$$

with

$$\alpha(\omega_0) = \frac{e^2/m}{\omega_s^2 - \omega_0^2 + i\omega_0\gamma} \tag{1-41}$$

A more precise evaluation of the *polarizability* $\alpha(\omega_0)$ can be given by means of a semiclassical approach (see, for example, Eyring *et al.*, 1944, Chapter VIII) in which only the molecular system is treated quantum mechanically since no relevant effect appears due to the quantization of the electromagnetic field. This treatment accounts explicitly for the molecular structure, giving rise, in particular, to a polarizability tensor $\alpha(\omega_0)$ for nonisotropic molecules.

Let us remember that the nonrelativistic Hamiltonian H for a system of charges q_i in the presence of an electromagnetic field is given by

$$H(t) = \sum_i \left\{ \frac{1}{2m_i}\left[\mathbf{p}_i - \frac{q_i}{c}\,\mathbf{A}(\mathbf{r}_i, t)\right]^2 + q_i\,\Phi(\mathbf{r}_i, t) \right\} + \tilde{V}(\{\mathbf{r}_j\}) \tag{1-42}$$

where $\mathbf{A}$ and Φ are the vector and the scalar potentials related to the electric and magnetic field $\mathbf{E}$ and $\mathbf{H}$, respectively, by the usual relations (Stratton, 1941, Chapter 1):

$$\mathbf{E} = -\frac{1}{c}\frac{\partial \mathbf{A}}{\partial t} - \nabla\Phi \tag{1-43}$$

$$\mathbf{H} = \nabla \times \mathbf{A}$$

$\mathbf{r}_i$ is the position of the ith charge at time t, $\mathbf{p}_i$ its conjugate momentum, and $\tilde{V}(\{\mathbf{r}_j\})$ represents the mutual interaction.

In a charge-free region the electromagnetic field can be represented by means of the vector potential $\mathbf{A}$ alone, since Φ can be set identically equal to zero. Furthermore, since the system of charges is a molecule for which all but a few electrons (optical electrons) are strongly bounded to the central nucleus, the influence of the electromagnetic perturbation is limited to these few electrons. Let us assume, for the sake of simplicity, that only one optical electron is present, so that the Hamiltonian reduces to

$$H(t) = \frac{1}{2m}\left[\mathbf{p} - \frac{e}{c}\mathbf{A}(\mathbf{r}, t)\right]^2 + V(\mathbf{r}) + \tilde{H} \qquad (1\text{-}44)$$

where $\mathbf{r}$ is the electron position, and $V(\mathbf{r})$ the potential exerted on it by the remaining charges, whose Hamiltonian is indicated by $\tilde{H}$. Accordingly, the time-dependent Schrödinger equation for the state $|\Psi(t)\rangle$ of the electron is

$$\frac{1}{2m}\left[i\hbar\nabla + \frac{e}{c}\mathbf{A}(\mathbf{r}, t)\right]^2|\Psi(t)\rangle + V(\mathbf{r})|\Psi(t)\rangle = i\hbar\frac{\partial}{\partial t}|\Psi(t)\rangle \quad (1\text{-}45)$$

A convenient approximation can be introduced whenever the wavelength of the radiation field is much greater than the dimension of the molecule. In this case one can approximate the vector potential by its value in the center of mass of the molecule $\mathbf{R}$ (supposed at rest), thus rewriting Eq. (1-45) as

$$\frac{1}{2m}\left[i\hbar\nabla + \frac{e}{c}\mathbf{A}(\mathbf{R}, t)\right]^2|\Psi(t)\rangle + V(\mathbf{r})|\Psi(t)\rangle = i\hbar\frac{\partial}{\partial t}|\Psi(t)\rangle \quad (1\text{-}46)$$

If one now performs the unitary transformation (Scully, 1969)

$$|\Psi(t)\rangle = \exp\left[-\frac{ie}{\hbar c}\mathbf{A}(\mathbf{R}, t)\cdot\mathbf{r}\right]|\psi(t)\rangle \qquad (1\text{-}47)$$

Eq. (1-46) becomes

$$\left[\frac{\mathbf{p}^2}{2m} + V(\mathbf{r}) + \frac{e}{c} \frac{\partial \mathbf{A}(\mathbf{R}, t)}{\partial t} \cdot \mathbf{r} \right] |\psi(t)\rangle$$

$$= [H_0 - e\mathbf{E}(\mathbf{R}, t) \cdot \mathbf{r}] |\psi(t)\rangle = i\hbar \frac{\partial}{\partial t} |\psi(t)\rangle \quad (1\text{-}48)$$

where H_0 is the Hamiltonian in the absence of the electromagnetic field, and use has been made of the relation $\mathbf{E} = -(1/c) \partial \mathbf{A}/\partial t$.

To avoid confusion, it is worth stressing that we use here the real representation for the electromagnetic field, in contrast with the complex representation used in the previous section. This assures that the transformation given in Eq. (1-47) will be unitary.

The evaluation of the quantum expectation value of any operator that commutes with the quantity $\exp[-(ie/\hbar c)\mathbf{A}(\mathbf{R}, t) \cdot \mathbf{r}]$ can be done by the simple substitution $|\Psi\rangle \rightarrow |\psi\rangle$, which is equivalent to using Eq. (1-48) instead of Eq. (1-45). This procedure is known as *dipole approximation*, since the term $-e\mathbf{E}(\mathbf{R}, t) \cdot \mathbf{r}$ is the interaction energy with the field $\mathbf{E}$ of a dipole $e\mathbf{r}$ placed at $\mathbf{R}$. Thus the quantum expectation value $\mathbf{P}$ of the dipole operator $e\mathbf{r}$ is given by

$$\mathbf{P} = \langle \psi | e\mathbf{r} | \psi \rangle \quad (1\text{-}49)$$

which can be evaluated by applying to Eq. (1-48) ordinary first-order, time-dependent perturbation theory. The final result is

$$\mathbf{P}(t) = \int_{-\infty}^{+\infty} \boldsymbol{\alpha}(\omega) \cdot \mathbf{E}_\omega(\mathbf{R}) e^{-i\omega t} d\omega \quad (1\text{-}50)$$

in the general case of a nonmonochromatic incident field

$$\mathbf{E}(\mathbf{R}, t) = \int_{-\infty}^{+\infty} \mathbf{E}_\omega(\mathbf{R}) e^{-i\omega t} d\omega \quad (1\text{-}51)$$

where the *polarizability tensor* is obtained in the form

$$\boldsymbol{\alpha}(\omega) = \frac{2}{\hbar} \sum_b \frac{\omega_{bg}}{\omega_{bg}^2 - \omega^2} \langle u_g^{(0)} | e\mathbf{r} | u_b^{(0)} \rangle \langle u_b^{(0)} | e\mathbf{r} | u_g^{(0)} \rangle \quad (1\text{-}52)$$

where the $|u_b^{(0)}\rangle$ and the E_b are the eigenvectors and the eigenvalues of the unperturbed Hamiltonian H_0, the symbol g labels the ground state, and $\omega_{bg} = (E_b - E_g)/\hbar$ [the divergence of $\alpha(\omega)$ for $\omega = \omega_{bg}$ can be removed by considering the finite lifetime of the excited states, which has its classical counterpart in the radiation damping term $\gamma \, d\mathbf{r}/dt$ appearing in Eq. (1-34)].

The polarizability α reduces to a scalar quantity α for nonpolar isotropic molecules, that is, when the electron cloud does not possess a preferential orientation, so that

$$\alpha(\omega) = \frac{2}{3\hbar} \sum_b \frac{\omega_{bg}}{\omega_{bg}^2 - \omega^2} |\langle u_g^{(0)}| e\mathbf{r} |u_b^{(0)}\rangle|^2$$

where the factor $\frac{1}{3}$ derives from the expectation value over the equiprobable orientations of the molecule.

We shall limit ourselves to consideration of systems composed of optically isotropic molecules. If $\alpha(\omega)$ does not vary appreciably over the bandwidth of $\mathbf{E}(\mathbf{R}, t)$, which is supposed to be centered around the frequency ω_0, one can rewrite Eq. (1-50) as

$$\mathbf{P}(t) = \alpha(\omega_0)\,\mathbf{E}(\mathbf{R}, t) \qquad (1\text{-}53)$$

The preceding calculation, which is strictly valid for a molecule at rest, can be easily generalized to the situation in which the center of mass is moving along a trajectory $\mathbf{R}(t)$, such that $|\dot{\mathbf{R}}(t)| \ll c$. In this case one has

$$\mathbf{P}(t) = \alpha(\omega_0)\,\mathbf{E}[\mathbf{R}(t), t] \qquad (1\text{-}54)$$

if the function $\mathbf{E}(t) \equiv \mathbf{E}[\mathbf{R}(t), t]$ is supposed to possess a nearly monochromatic Fourier expansion centered around the frequency ω_0.

I.4 The Analytic Signal

The results of the preceding section have been found by adopting the real representation for the electromagnetic field. On the other hand, as we have seen in Section I.2, a complex representation can in some cases be useful for its formal advantages. This is particularly true for the *analytic representation*, which is peculiar for its close connection with the very nature of the photon detection process, as we shall see.

Let us deal with a real signal $X(t)$, which admits the Fourier expansion

$$X(t) = \int_{-\infty}^{+\infty} \xi(\omega) e^{-i\omega t} d\omega \tag{1-55}$$

Since $X(t)$ is a real function, one has

$$\xi(\omega) = \xi^*(-\omega) \tag{1-56}$$

that is, if we put $\xi(\omega) = \xi_1(\omega) + i\xi_2(\omega)$,

$$\xi_1(\omega) = \xi_1(-\omega)$$
$$\xi_2(\omega) = -\xi_2(-\omega) \tag{1-57}$$

which allows us to rewrite Eq. (1-55) as

$$X(t) = 2\int_0^{+\infty} [\xi_1(\omega)\cos\omega t + \xi_2(\omega)\sin\omega t]\, d\omega \tag{1-58}$$

The *analytic signal* $\hat{X}(t)$ is defined as (Born and Wolf 1970, Chapter X)

$$\hat{X}(t) = 2\int_0^{+\infty} \xi(\omega) e^{-i\omega t} d\omega \tag{1-59}$$

from which it follows that

$$\mathrm{Re}\,\hat{X}(t) = X(t) \tag{1-60}$$

where Re denotes the real part.

Along the same line, one can introduce the analytic signal $\hat{\mathbf{P}}(t)$, corresponding to the electric dipole moment. A word of warning is necessary, since it is not true in general that Eq. (1-54) generalizes to give

$$\hat{\mathbf{P}}(t) = \alpha(\omega_0)\hat{\mathbf{E}}[\mathbf{R}(t), t] \tag{1-61}$$

where $\hat{\mathbf{E}}[\mathbf{R}(t), t]$ is the analytic electric field $\hat{\mathbf{E}}[\mathbf{R}, t]$ evaluated at the space–time point $[\mathbf{R}(t), t]$. Nevertheless, it is easy to show that Eq. (1-61) can be considered to be valid under the assumption $|\dot{\mathbf{R}}(t)| \ll c$, which is the nonrelativistic hypothesis underlying the entire development. We shall consider only the case in which $\mathbf{E}(\mathbf{R}, t)$ is a monochromatic field:

$$\mathbf{E}(\mathbf{R}, t) = \mathbf{E}_0 \cos(\mathbf{k}_0 \cdot \mathbf{R} - \omega_0 t) \qquad (1\text{-}62)$$

but the demonstration can be easily generalized to a narrow bandwidth field. After introducing

$$\mathbf{E}(t) = \mathbf{E}_0 \cos[\mathbf{k}_0 \cdot \mathbf{R}(t) - \omega_0 t] \qquad (1\text{-}63)$$

and

$$\exp[i\mathbf{k}_0 \cdot \mathbf{R}(t)] = \int_{-\infty}^{+\infty} \Psi(\omega) e^{-i\omega t} d\omega \qquad (1\text{-}64)$$

it follows that

$$\hat{\mathbf{E}}(t) = \mathbf{E}_0 \int_0^{+\infty} \Psi(\omega - \omega_0) e^{-i\omega t} d\omega$$

$$+ \mathbf{E}_0 \int_0^{+\infty} \Psi^*(-\omega - \omega_0) e^{-i\omega t} d\omega \qquad (1\text{-}65)$$

Since $\Psi(\Omega)$ is practically zero for $|\Omega| > k_0 v$, where v is an upper limit for the molecular velocity, Eq. (1-65) reduces approximately to

$$\hat{\mathbf{E}}(t) = \mathbf{E}_0 \int_{-\infty}^{+\infty} \Psi(\omega - \omega_0) e^{-i\omega t} d\omega = \mathbf{E}_0 \exp[i\mathbf{k}_0 \cdot \mathbf{R}(t) - i\omega_0 t]$$

$$(1\text{-}66)$$

whenever $k_0 v / \omega_0 = v/c \ll 1$. Observing that

$$\hat{\mathbf{E}}(\mathbf{R}, t) = \mathbf{E}_0 \exp[i(\mathbf{k}_0 \cdot \mathbf{R} - \omega_0 t)] \qquad (1\text{-}67)$$

follows from Eq. (1-62), one immediately obtains Eq. (1-61) from Eq. (1-66).

In a similar way, it is possible to demonstrate that, when considering the field generated by a moving dipole, the complex field obtained by writing $\hat{\mathbf{P}}$ instead of $\mathbf{P}$ in all equations represents the analytic field, under the same nonrelativistic approximation for the dipole motion. These considerations imply that once the incident field is represented in its analytic form, the induced dipole moments and electric fields that they generate are directly obtained in their analytic form.

I.5 The Effective Dipole Moment

The electric field acting on a molecule consists not only of the external radiation field (that is, the field that would interact with the supposedly isolated molecule), but also includes the perturbation due to the other molecules present in the system, which is in turn associated with their induced electric dipole moment. Therefore Eq. (1-61) has to be used, keeping in mind that $\hat{\mathbf{E}}[\mathbf{R}(t), t]$ is the *effective* field at point $\mathbf{R}(t)$ at time t. Since this effective field is a functional of the induced dipole moments, these can be evaluated by means of a system of coupled equations obtained by applying Eq. (1-61) to each molecule.

Let us assume that the external electromagnetic radiation acting upon our system is a plane polarized monochromatic wave of the form

$$\hat{\mathbf{E}}_{ext}(\mathbf{r}, t) = \mathbf{E}_0 \exp[i(\mathbf{k}_0 \cdot \mathbf{r} - \omega_0 t)] \qquad (1\text{-}68)$$

Under the influence of the external field alone, the dipole moment associated with the ith molecule of trajectory $\mathbf{R}_i(t)$ is, according to Eq. (1-61),

$$\hat{\mathbf{P}}_i(t) = \alpha(\omega_0) \mathbf{E}_0 \exp\{i[\mathbf{k}_0 \cdot \mathbf{R}_i(t) - \omega_0 t]\} \qquad (1\text{-}69)$$

The exact expression for $\hat{\mathbf{P}}_i(t)$ is obtained by simply adding to the external field given by Eq. (1-68) the microfields radiated by all the other molecules. This is justified by the nonrelativistic assumption, which ensures that all the microfields are still quasimonochromatic with frequency ω_0, thus implying the validity of Eq. (1-61). This amounts to saying that the dipole moment has a quasimonochromatic behavior with frequency ω_0 and width $\Delta\omega$ such that $\Delta\omega/\omega_0 \leqslant v/c \ll 1$. In order to evaluate the electric field radiated by such a dipole, let us first consider the monochromatic case, where

$$\hat{\mathbf{P}}_i(t) = \mathbf{P}_{0i}\, e^{-i\omega_0 t} \tag{1-70}$$

We remember that, according to Eq. (1-18), the Hertz vector at point $\mathbf{r}$ and time t radiated by a molecule at rest is given by

$$\hat{\boldsymbol{\Pi}}^{(0)}(\mathbf{r},t) = \frac{\hat{\mathbf{P}}[t-(r/c)]}{r} \tag{1-71}$$

where the origin of the reference frame is assumed to coincide with the center of the molecule. The generalization of Eq. (1-71) to the case in which the ith molecule moves along the trajectory $\mathbf{R}_i(t)$ is

$$\hat{\boldsymbol{\Pi}}_i^{(0)}(\mathbf{r},t) = \int \frac{\hat{\mathbf{P}}_i[t-(|\mathbf{r}'-\mathbf{r}|/c)]}{|\mathbf{r}-\mathbf{r}'|} \delta\left[\mathbf{r}' - \mathbf{R}_i\left(t - \frac{|\mathbf{r}'-\mathbf{r}|}{c}\right)\right] d\mathbf{r}' \tag{1-72}$$

which means that the field at the space–time point $(\mathbf{r},t)$ is generated at the point $\mathbf{r}'$ such that a light signal emitted by the molecule in passing $\mathbf{r}'$ reaches the point $\mathbf{r}$ at time t. According to Eq. (1-70), one can rewrite Eq. (1-72) in the form

$$\hat{\boldsymbol{\Pi}}_i^{(0)}(\mathbf{r},t) = \int \frac{\mathbf{P}_{0i}\exp\left\{-i\omega_0\left[t - \frac{|\mathbf{r}-\mathbf{R}_i[t-(|\mathbf{r}'-\mathbf{r}|/c)]|}{c}\right]\right\}}{|\mathbf{R}_i[t-(|\mathbf{r}'-\mathbf{r}|/c)] - \mathbf{r}|}$$
$$\times\, \delta\left[\mathbf{r}' - \mathbf{R}_i(t - \frac{|\mathbf{r}'-\mathbf{r}|}{c})\right] d\mathbf{r}' \tag{1-73}$$

This equation can be simplified if, according to our purposes of evaluating the field generated by one molecule on the others, we assume that the point $\mathbf{r}$ lies inside the scattering volume. In this case one can approximate by writing

$$\mathbf{R}_i\left(t - \frac{|\mathbf{r}'-\mathbf{r}|}{c}\right) \simeq \mathbf{R}_i(t) - \frac{\dot{\mathbf{R}}_i(t)}{c}|\mathbf{r}'-\mathbf{r}| \tag{1-74}$$

from which it follows that on the right-hand side of Eq. (1-73) one has

$$
\begin{aligned}
|\mathbf{R}_i(t) - \mathbf{r}| &= \left| \mathbf{R}_i\left(t - \frac{|\mathbf{r}' - \mathbf{r}|}{c} \right) + \frac{\dot{\mathbf{R}}_i(t)}{c} |\mathbf{r}' - \mathbf{r}| - \mathbf{r} \right| \\
&= \left| \mathbf{r}' + \frac{\dot{\mathbf{R}}_i(t)}{c} |\mathbf{r}' - \mathbf{r}| - \mathbf{r} \right| \\
&= |\mathbf{r}' - \mathbf{r}| \left[1 + O\left(\frac{v}{c} \right) \right]
\end{aligned}
\tag{1-75}
$$

where O is the usual order symbol. The further condition $(v/c)(L/\lambda_0) \ll 1$ (L being the length of the path the molecule can travel, that is, in practice, the linear dimension of the scattering volume) allows one, with the help of Eq. (1-75), to rewrite Eq. (1-73) in the form

$$
\hat{\mathbf{\Pi}}_i^{(0)}(\mathbf{r}, t) = \mathbf{P}_{0i} \frac{1}{|\mathbf{R}_i(t) - \mathbf{r}|} \exp\left\{ -i\omega_0 \left[t - \frac{|\mathbf{r} - \mathbf{R}_i(t)|}{c} \right] \right\}
\tag{1-76}
$$

The electric field generated by the ith molecule at point $\mathbf{r}$ is, according to Eq. (1-11), then given by

$$
\hat{\mathbf{E}}_i(\mathbf{r}, t) = (\nabla\nabla + k_0^2 \, \mathbf{U}) \cdot \mathbf{P}_{0i} e^{-i\omega_0 t} \frac{\exp[ik_0|\mathbf{r} - \mathbf{R}_i(t)|]}{|\mathbf{r} - \mathbf{R}_i(t)|}
\tag{1-77}
$$

where $\mathbf{U}$ is the unit tensor, and use has been made of the approximate relation $\partial^2 \hat{\mathbf{\Pi}}_i^{(0)}/\partial t^2 = -\omega_0^2 \, \hat{\mathbf{\Pi}}_i^{(0)}$. This expression can be easily generalized to the case of a quasimonochromatic dipole

$$
\hat{\mathbf{P}}_i(t) = \int_0^{+\infty} \mathbf{P}_{i\omega} e^{-i\omega t} \, d\omega
\tag{1-78}
$$

centered around $\omega = \omega_0$. In fact, Eq. (1-77) immediately yields

$$
\hat{\mathbf{E}}_i(\mathbf{r}, t) = \int_0^{+\infty} (\nabla\nabla + k^2 \, \mathbf{U}) \cdot \mathbf{P}_{i\omega} e^{-i\omega t} \frac{\exp[ik|\mathbf{r} - \mathbf{R}_i(t)|]}{|\mathbf{r} - \mathbf{R}_i(t)|} \, d\omega
\tag{1-79}
$$

where $k = \omega/c$, which, under the already assumed condition

$(v/c)(L/\lambda_0) \ll 1$, reduces to

$$\hat{\mathbf{E}}_i(\mathbf{r}, t) = (\nabla\nabla + k_0^2 \,\mathbf{U}) \cdot \hat{\mathbf{P}}_i(t) \frac{\exp[ik_0|\mathbf{r} - \mathbf{R}_i(t)|]}{|\mathbf{r} - \mathbf{R}_i(t)|} \qquad (1\text{-}80)$$

The relation between the dipole moment $\hat{\mathbf{P}}_i$ and the associated radiation field $\hat{\mathbf{E}}_i$ allows us to express in a self-consistent way the fact that each molecule is under the influence of the external field $\hat{\mathbf{E}}_{\text{ext}}$ as well as of the microfields generated by all the other molecules, which results in writing a system of coupled equations for the induced dipole moments. However, one has to keep in mind that the validity of the previous considerations relies on the dipole approximation, that is, on the assumption that the field seen by each molecule does not vary appreciably in its linear dimension. This is certainly the case for the external field, but it is not true for fields generated on the ith molecule by the neighboring ones, that is, by those whose distance from it is of the order of a few molecular diameters. Since the magnitude of each of these fields is far from negligible, an exact calculation should take into account their contributions, resulting in a more rigorous procedure than the one furnished by the dipole approximation. In any event, the total effect of the molecules contained in a sphere whose radius is small compared with the wavelength centered around the position of the ith molecule tends to be negligible due to the particular form of the electromagnetic field radiated by identical dipoles. This fact can be rigorously proved if the molecules are thought to occupy completely symmetrical positions around the center of the sphere (Lorentz, 1916, Chapter IV; Becker, 1933, Chapter III), a condition met with good approximation for condensed systems. On the other hand, the density of a rarefied gas for which the previous argument does not apply is small enough so that we can neglect the electromagnetic field radiated by molecules that happen to be at short distances from the ith dipole.

These considerations ensure the existence of a large class of material media for which the influence of the near molecules can be neglected in the evaluation of the induced dipole moment. Therefore Eq. (1-61) yields for the dipole moment of the jth molecule

$$\hat{\mathbf{P}}_j(t) = \alpha(\omega_0) \left\{ \hat{\mathbf{E}}_{\text{ext}}[\mathbf{R}_j(t), t] + \sum_{i \neq j}' \hat{\mathbf{E}}_i[\mathbf{R}_j(t), t] \right\} \quad (1\text{-}81)$$

where $\mathbf{R}_j(t)$ labels the position of the center of the jth molecule and the primed sum is extended to the molecules outside the small sphere under consideration. In practice, one can extend the sum to all the molecules, since the total perturbation due to the dipoles inside the sphere is in any case negligible. By taking into account Eqs. (1-68) and (1-80), a closed system of equations is obtained in the form

$$\hat{\mathbf{P}}_j(t) = \alpha(\omega_0) \left\{ \mathbf{E}_0 \exp[i(\mathbf{k}_0 \cdot \mathbf{R}_j(t) - \omega_0 t)] + \sum_{i \neq j}^{1, N} \mathbf{T}_{ji} \cdot \hat{\mathbf{P}}_i(t) \right\}$$
$$(1\text{-}82)$$

where N is the total number of molecules acted on by the external field, and the tensor operator $\mathbf{T}_{ji}$ is defined by

$$\mathbf{T}_{ji} \equiv (\nabla_j \nabla_j + k_0^2 \mathbf{U}) \exp[ik_0 s_{ji}(t)]/s_{ji}(t) \quad (1\text{-}83)$$

where $s_{ji}(t) = |\mathbf{R}_j(t) - \mathbf{R}_i(t)|$, and the index j in ∇_j means that the gradient has to be evaluated with respect to the coordinate $\mathbf{R}_j$.

The set of N equations (1-82) represents the dynamical generalization of the classical integral equations for the dipole moments, valid in the static case, that is, when the molecules are assumed to have fixed positions in time. The static case has been extensively treated in the literature, and a detailed analysis can be found in many texts (Born and Wolf, 1970, Chapter II; Rosenfeld, 1951, Chapter VI).

I.6 The Scattered Field

The microfields generated by the dipoles $\hat{\mathbf{P}}_j$ associated with each molecule combine to produce the scattered electromagnetic field, actually observed outside the region occupied by the material system. Therefore, the evaluation of the scattered field $\hat{\mathbf{E}}_{sc}$ requires, as a first step, knowledge of the value of the $\hat{\mathbf{P}}_j$'s obtained by solving Eq. (1-82). This can be done in principle by means of an iterative procedure, which yields

$$\hat{\mathbf{P}}_j(t) = \alpha(\omega_0)\,\mathbf{E}_0\,\exp[i\mathbf{k}_0 \cdot \mathbf{R}_j(t) - i\omega_0 t]$$

$$+ \alpha^2(\omega_0)\sum_{\substack{i \mp j}}^{1,N} \mathbf{T}_{ji} \cdot \mathbf{E}_0\,\exp[i\mathbf{k}_0 \cdot \mathbf{R}_i(t) - i\omega_0 t]$$

$$+ \alpha^3(\omega_0)\sum_{\substack{i \mp j}}^{1,N}\sum_{\substack{l \mp i}}^{1,N} \mathbf{T}_{ji} \cdot (\mathbf{T}_{il} \cdot \mathbf{E}_0)\,\exp[i\mathbf{k}_0 \cdot \mathbf{R}_l(t) - i\omega_0 t] + \cdots$$

$$(1\text{-}84)$$

where the first term on the right-hand side represents the direct contribution of the external field and is independent of the other molecules present in the system. Their influence is shown by the successive terms, which can be interpreted as *multiple scattering* contributions, since they are related to the field acting on the *j*th molecule after successive interactions with other molecules. More explicitly, the term involving the quantity $\mathbf{T}_{ji}$ represents the influence of the external field after scattering by the *i*th molecule, the one containing $\mathbf{T}_{ji}\,\mathbf{T}_{il}$ indicates a double scattering by the *i*th and *l*th molecule, and so on.

The scattered radiation can now be obtained by substituting Eq. (1-84) into Eq. (1-72). In this way, $\hat{\mathbf{E}}_{sc}$ will result in the sum of various terms, having the meaning of single, double, triple, etc., scattering. Multiple scattering plays a relevant role when considering the fundamental problem of the connection between microscopic and macroscopic theories of light scattering as will be discussed in detail in Chapter III. We now evaluate the scattered field taking into account only the first term of Eq. (1-84). To this end, we apply Eq. (1-72), keeping in mind that in an actual scattering experiment the observation point is usually placed at a distance from the scattering volume large compared to its linear dimension, which amounts to saying that

$r \gg r'$ if the origin of the reference frame is placed inside the scattering volume. Under this condition, Eqs. (1-72) and (1-84) yield

$$\hat{\Pi}_{sc}(\mathbf{r}, t) = \sum_{j=1}^{N} \hat{\Pi}_{jsc}^{(0)}(\mathbf{r}, t)$$

$$= \frac{\alpha(\omega_0) \mathbf{E}_0}{r} \sum_{j=1}^{N} \int \exp\left[i\mathbf{k}_0 \cdot \mathbf{R}_j\left(t - \frac{|\mathbf{r}' - \mathbf{r}|}{c}\right)\right]$$

$$\times \exp\left[-i\omega_0 t + \frac{i\omega_0 r}{c} - \frac{i\omega_0 \mathbf{r}' \cdot \mathbf{\eta}}{c}\right]$$

$$\times \delta\left[\mathbf{r}' - \mathbf{R}_j\left(t - \frac{|\mathbf{r}' - \mathbf{r}|}{c}\right)\right] d\mathbf{r}' \qquad (1\text{-}85)$$

having taken advantage of the fact that for large r one can approximate by writing

$$|\mathbf{r}' - \mathbf{r}| \simeq r - \mathbf{r}' \cdot \mathbf{\eta} \qquad (1\text{-}86)$$

with $\mathbf{\eta} = \mathbf{r}/r$. The property of the δ-function allows us to rewrite Eq. (1-85) in the form

$$\hat{\Pi}_{sc}(\mathbf{r}, t) = \frac{\alpha(\omega_0) \mathbf{E}_0}{r} \sum_{j=1}^{N} \int \exp\left[i\left(\mathbf{k}_0 - \frac{\omega_0}{c}\mathbf{\eta}\right) \cdot \mathbf{R}_j\left(t - \frac{|\mathbf{r}' - \mathbf{r}|}{c}\right)\right]$$

$$\times \exp\left[-i\omega_0\left(t - \frac{r}{c}\right)\right] \delta\left[\mathbf{r}' - \mathbf{R}_j\left(t - \frac{|\mathbf{r}' - \mathbf{r}|}{c}\right)\right] d\mathbf{r}'$$

$$(1\text{-}87)$$

We now observe that the approximation $\mathbf{R}_j[t - (|\mathbf{r}' - \mathbf{r}|/c)] \simeq \mathbf{R}_j[t - (r/c)]$ implies in the exponential appearing in Eq. (1-87) an error of the order of $(v/c)(L/\lambda_0)$, a quantity that has been assumed in the previous section to be much less than one.[†]

[†] We wish to emphasize that the condition $(v/c)(L/\lambda_0) \ll 1$ is necessary for the validity of Eq. (1-84), as far as multiple scattering is concerned. Whenever only single scattering is present this condition simply allows us to rewrite Eq. (1-87) in the compact form of Eq. (1-88).

Therefore, Eq. (1-87) can be cast in the form

$$\hat{\mathbf{\Pi}}_{sc}(\mathbf{r}, t) = \frac{\alpha(\omega_0)\,\mathbf{E}_0}{r} \sum_{j=1}^{N} \exp\left[i\mathbf{k}_1 \cdot \mathbf{R}_j\left(t - \frac{r}{c}\right) - i\omega_0\left(t - \frac{r}{c}\right)\right] \tag{1-88}$$

where $\mathbf{k}_1 = \mathbf{k}_0 - k_0\,\mathbf{\eta}$ (see Fig. 1.1). According to Eq. (1-11) the expression for $\hat{\mathbf{E}}_{sc}$ can easily be derived from Eq. (1-88), taking into account the vector identity $\nabla(\nabla \cdot \mathbf{a}) = \nabla^2\mathbf{a} + \nabla \times (\nabla \times \mathbf{a})$ and the already used nonrelativistic hypothesis, in the form (Komarov and Fisher, 1962)

$$\hat{\mathbf{E}}_{sc}(\mathbf{r}, t) = -\frac{\alpha(\omega_0)\,k_0^2}{r}\,\mathbf{\eta} \times (\mathbf{\eta} \times \mathbf{E}_0) \sum_{j=1}^{N} \exp\left[i\mathbf{k}_1 \cdot \mathbf{R}_j\left(t - \frac{r}{c}\right)\right]$$

$$\times \exp\left[-i\omega_0\left(t - \frac{r}{c}\right)\right] \tag{1-89}$$

The treatment leading to Eq. (1-89) is completely classical, apart from the evaluation of the polarizability. In any event, the expression of the scattered spectrum, which can be obtained by means of Eq. (1-89) [see Eq. (2-50)], has the same form as that furnished by a fully quantum-mechanical treatment (Gelbart, 1974). A quantum-mechanical expression for the scattered field itself has been obtained by Hellwarth (1970). This author points out the relevance of the quantum approach whenever the frequency shifts associated with the scattering process times Planck's constant $\hbar$ are not very much less than $K_B T$ (Boltzmann's constant times absolute temperature).

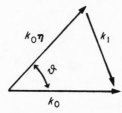

Fig. 1.1 *Vector relation pertinent to scattering.*

II

The Scattered Field as a Random Variable

II.1 Ensemble Averages

The preceding development of microscopic molecular scattering theory refers to a case in which all the variables involved are considered well-prescribed functions of time. In this way, the scattered electric field also turns out to be a prescribed function of time, whose value depends on the positions of the scattering molecules as well as on the incident radiation field. One has, however, to keep in mind that such detailed information is not generally accessible by actual experiment and is not always meaningful. As a matter of fact, in the optical frequency range the time response of the available detectors is such that a time average over many periods of oscillation of the field is automatically performed. Furthermore, the kind of information on the scattering medium one usually wishes to extract

27

from a scattering experiment does not concern the motion of the single molecule, but the statistical behavior of the system as a whole. Thus one has to manipulate the expression giving the deterministic value of the scattered field in order to obtain quantities that can be compared with those actually measured. This process will depend on the kind of experiment one is performing, the type of detectors used, etc., so that a precise choice can be made only case by case. This procedure, which one has to employ when comparing the theory with the results of the experiment, is obviously not very general. One may prefer then to describe the scattered field in terms of *ensemble averages*. This is a completely general way of characterizing a random variable by means of averages, which in principle must be performed by repeating the experiment many times on different samples of the same system specified by different initial conditions. These averages may or may not have a direct connection with the particular quantities that are found in a particular sample. For example, one can identify these ensemble averages with time averages in the case of ergodic processes, but this is never true for nonstationary situations. In this case, the time-averaging operation furnishes no meaningful information, and ensemble averaging represents the only way of describing the physical behavior of the system.

It is worth stressing that, from a theoretical point of view, ensemble averages can be evaluated in most cases, while time averages cannot, since this would imply the possibility of following the evolution of the system for every single realization.

The ensemble-averaging approach will be adopted in the following to study the scattered field, considering situations in which all fluctuations are due entirely to a statistical uncertainty associated with the translational motion of the molecules, i.e., ignoring cases in which the nature of the incident radiation field and the randomness of molecular orientation[†]

[†]Obviously, the randomness of orientation has no influence on the scattered field if the polarizability is a scalar, that is, for optically isotropic molecules.

are further sources of fluctuations. To this end, our first step will be the introduction of the hierarchy of correlation functions describing the statistical properties of the scattering system.

II.2 The N-Particle Distribution Function

We consider a system consisting of N identical molecules whose *microscopic* state is specified at any given instant by a set of $6N$ conjugate variables $\{\mathbf{q}_i, \mathbf{p}_i\}$ representing the position and momentum of all particles. According to the adopted ensemble-averaging approach, we introduce an ensemble consisting of many replica systems, each of which is characterized by the values $\{\mathbf{q}_{i0}, \mathbf{p}_{i0}\}$ assumed by the conjugate variables at time $t = 0$. The most complete way of describing this ensemble is to consider the N-particle *distribution function* $F_N(\{\mathbf{q}_i, \mathbf{p}_i\})$, i.e., the relative density of systems in the state $\{\mathbf{q}_i, \mathbf{p}_i\}$ (Landau and Lifshitz, 1958, Chapter I). This procedure can be formalized by introducing the $6N$-dimensional space Γ whose volume element is defined as $d\Gamma = \prod_{i=1}^{N} d\mathbf{q}_i \, d\mathbf{p}_i$ so that the relative number of systems whose coordinates range in a small volume $d\Gamma$ centered around the point $\{\mathbf{q}_i, \mathbf{p}_i\}$ is given by

$$F_N(\{\mathbf{q}_i, \mathbf{p}_i\}) \prod_{i=1}^{N} d\mathbf{q}_i \, d\mathbf{p}_i \tag{2-1}$$

where the function F_N plays the role of normalized density in the Γ-space, since it is assumed to fulfill the condition

$$\int F_N(\{\mathbf{q}_i, \mathbf{p}_i\}) \prod_{i=1}^{N} d\mathbf{q}_i \, d\mathbf{p}_i = 1 \tag{2-2}$$

The function F_N is obviously symmetrical under the exchange of any two particles, due to the fact that the state of the system is not altered by this operation. It allows us to evaluate the average value of any physical quantity $G(\{\mathbf{q}_i, \mathbf{p}_i\})$ depending on the microscopic state $\{\mathbf{q}_i, \mathbf{p}_i\}$ of the system. As a matter of fact, the definition of F_N ensures that

$$\langle G \rangle = \int G(\{\mathbf{q}_i, \mathbf{p}_i\}) F_N(\{\mathbf{q}_i, \mathbf{p}_i\}) \prod_{i=1}^{N} d\mathbf{q}_i \, d\mathbf{p}_i \qquad (2\text{-}3)$$

where $\langle G \rangle$ is the *average value* of G over the ensemble of identical systems under consideration. The temporal behavior of the distribution function F_N is expressed through the *Liouville equation* (*first Liouville theorem*) (Landau and Lifshitz, 1958, Chapter I)

$$\frac{dF_N}{dt}(\{\mathbf{q}_i, \mathbf{p}_i\}, t) = \sum_{i=1}^{N} \left(\frac{\partial F_N}{\partial \mathbf{q}_i} \cdot \dot{\mathbf{q}}_i + \frac{\partial F_N}{\partial \mathbf{p}_i} \cdot \dot{\mathbf{p}}_i \right) + \frac{\partial F_N}{\partial t} = 0 \qquad (2\text{-}4)$$

which is a consequence of the conservation of the total number of systems and amounts to saying that the distribution function is constant when evaluated along the trajectory followed by a point $\{\mathbf{q}_i, \mathbf{p}_i\}$ representing a given system in Γ-space. It is worthwhile remembering that Eq. (2-4) is usually associated with the *second Liouville theorem* (Sommerfeld, 1956, Chapter IV), which states that each volume element $\Delta\Gamma$ formed by the points representing a given number of well-identified systems in Γ-space maintains a constant value with time, that is,

$$\frac{d}{dt} \Delta\Gamma = 0 \qquad (2\text{-}5)$$

The use of the distribution function furnishes a complete description of the temporal evolution relative to an ensemble of systems, in the sense that F_N contains the positions and momenta of all N particles constituting each system. This completeness corresponds to the practical impossibility of evaluating F_N in many cases of interest. On the other hand, one is more often interested in physical quantities depending on a very small fraction of the $6N$ conjugate variables $\{\mathbf{q}_i, \mathbf{p}_i\}$, so that the complete knowledge furnished by F_N is hardly necessary. Therefore, the problem arises of looking for descriptions that, although less complete, give rise to equations that can be handled in actual situations.

II.3 The Hierarchy of Reduced Distribution Functions

Let us consider a quantity $G(\mathbf{q}_1, ..., \mathbf{q}_s, \mathbf{p}_1, ..., \mathbf{p}_s)$ depending only on $6s$ conjugate variables, with $s < N$. In this case, Eq. (2-3) becomes

$$
\begin{aligned}
\langle G \rangle &= \int G(\mathbf{q}_1, ..., \mathbf{q}_s, \mathbf{p}_1, ..., \mathbf{p}_s) \prod_{i=1}^{s} d\mathbf{q}_i \, d\mathbf{p}_i \\
&\quad \times \int F_N(\{\mathbf{q}_i, \mathbf{p}_i\}) \prod_{i=s+1}^{N} d\mathbf{q}_i \, d\mathbf{p}_i \\
&= \int G(\mathbf{q}_1, ..., \mathbf{q}_s, \mathbf{p}_1, ..., \mathbf{p}_s) \, F_s(\mathbf{q}_1, ..., \mathbf{q}_s, \mathbf{p}_1, ..., \mathbf{p}_s) \\
&\quad \times \prod_{i=1}^{s} d\mathbf{q}_i \, d\mathbf{p}_i
\end{aligned} \tag{2-6}
$$

having defined the reduced s-particle distribution function as

$$
F_s(\mathbf{q}_1, ..., \mathbf{q}_s, \mathbf{p}_1, ..., \mathbf{p}_s) = \int F_N(\{\mathbf{q}_i, \mathbf{p}_i\}) \prod_{i=s+1}^{N} d\mathbf{q}_i \, d\mathbf{p}_i \tag{2-7}
$$

The function F_s, which is obviously symmetrical under the exchange of any two particles, plays the role of reduced normalized density in the Γ-space, so that it represents the probability of finding any s particles of each system in the small volume element $\Delta \Gamma_s = \prod_{i=1}^{s} d\mathbf{q}_i \, d\mathbf{p}_i$ around the point $(\mathbf{q}_1, ..., \mathbf{q}_s, \mathbf{p}_1, ..., \mathbf{p}_s)$.

We now observe that, if the considered ensemble is composed of systems whose particles are acted upon by forces not depending on velocity, we can use the ordinary coordinates $\{\mathbf{r}_i, m\mathbf{v}_i\}$ and then consider the $6N$-dimensional space Γ' obtained from Γ by means of the substitution $(m\mathbf{v}_1, ..., m\mathbf{v}_N) \rightarrow (\mathbf{v}_1, ..., \mathbf{v}_N)$. Since this is practically the case for our molecular system, it is worth rewriting the preceding relations between average values and correlation functions in the space Γ'. More precisely let us consider a physical quantity $G(\mathbf{r}_1, \ldots, \mathbf{r}_n, \mathbf{v}_1, \ldots, \mathbf{v}_n)$ $(n \leqslant N$, the equality sign representing the case of a quantity depending on all $6N$ coordinates). If we define the n-particle distribution function $F_n(\mathbf{r}_1, \ldots, \mathbf{r}_n, \mathbf{v}_1, \ldots, \mathbf{v}_n)$ as the probability of finding any n particles of each system in the small volume element $\Delta\Gamma'_n = \prod_{i=1}^{n} d\mathbf{r}_i \, dv_i$ around the point $(\mathbf{r}_1, \ldots, \mathbf{r}_n, \mathbf{v}_1, \ldots, \mathbf{v}_n)$, we have

$$\langle G(\mathbf{r}_1, \ldots, \mathbf{r}_n, \mathbf{v}_1, \ldots, \mathbf{v}_n) \rangle = \int G(\mathbf{r}_1, \ldots, \mathbf{r}_n, \mathbf{v}_1, \ldots, \mathbf{v}_n)$$

$$\times F_n(\mathbf{r}_1, \ldots, \mathbf{r}_n, \mathbf{v}_1, \ldots, \mathbf{v}_n) \prod_{i=1}^{n} d\mathbf{r}_i \, dv_i$$

$$(2\text{-}8)$$

where

$$F_n(\mathbf{r}_1, \ldots, \mathbf{r}_n, \mathbf{v}_1, \ldots, \mathbf{v}_n)$$

$$= \int F_N(\mathbf{r}_1, \ldots, \mathbf{r}_N, \mathbf{v}_1, \ldots, \mathbf{v}_N) \prod_{i=n+1}^{N} d\mathbf{r}_i \, dv_i, \qquad n < N \qquad (2\text{-}9)$$

the meaning of probability being assured by the normalization condition

$$\int F_N(\mathbf{r}_1, \ldots, \mathbf{r}_N, \mathbf{v}_1, \ldots, \mathbf{v}_N) \prod_{i=1}^{N} d\mathbf{r}_i \, dv_i = 1 \qquad (2\text{-}10)$$

which yields, with the help of Eq. (2-9), the normalization condition

$$\int F_n(\mathbf{r}_1, \ldots, \mathbf{r}_n, \mathbf{v}_1, \ldots, \mathbf{v}_n) \prod_{i=1}^{n} d\mathbf{r}_i \, dv_i = 1 \qquad (2\text{-}11)$$

for the reduced distribution functions. The Liouville equation obviously transforms into

$$\frac{dF_N}{dt}(\{\mathbf{r}_i, \mathbf{v}_i\}, t) = \sum_{i=1}^{N} \left(\frac{\partial F_N}{\partial \mathbf{r}_i} \cdot \dot{\mathbf{r}}_i + \frac{\partial F_N}{\partial \mathbf{v}_i} \cdot \dot{\mathbf{v}}_i \right) + \frac{\partial F_N}{\partial t} = 0 \qquad (2\text{-}12)$$

while the second Liouville theorem is

$$\frac{d\Delta\Gamma'}{dt} = 0 \tag{2-13}$$

The dynamical behavior of the reduced distribution functions F_n is described by a hierarchy of integrodifferential equations, which can be obtained by performing suitable integrations of Eq. (2-12) over volume elements of the kind $\prod_{i=n+1}^{N} d\mathbf{r}_i \, d\mathbf{v}_i$. One obtains in this way the well-known *Bogoliubov, Born, Green, Kirkwood, and Yvon (BBGKY) hierarchy*, which is (see, for example, Grad, 1958; Fisher, 1964, Chapter IV)

$$\frac{\partial F_n}{\partial t} + \sum_{i=1}^{n} \mathbf{v}_i \cdot \frac{\partial F_n}{\partial \mathbf{r}_i} + \sum_{i,j=1}^{n} \mathbf{K}_{ij} \cdot \frac{\partial F_n}{\partial \mathbf{v}_i}$$

$$+ (N-n) \sum_{i=1}^{n} \int \mathbf{K}_{i,n+1} \cdot \frac{\partial F_{n+1}}{\partial \mathbf{v}_i} \, d\mathbf{r}_{n+1} \, d\mathbf{v}_{n+1} = 0$$

$$n = 1, 2, \ldots, N \tag{2-14}$$

where $\mathbf{K}_{ij}$ is the acceleration of the ith particle due to the jth particle and $\mathbf{K}_{ii}$ the acceleration due to the external force. The system of Eqs. (2-14) is open, in the sense that the number of unknown functions is always one more than the number of equations. Therefore, some suitable approximation has to be introduced as a supplementary closure condition.

The simplest way of closing the system is to postulate the absence of any *correlation* between the particles, which corresponds to the condition

$$F_n(\mathbf{r}_1, \ldots, \mathbf{r}_n, \mathbf{v}_1, \ldots, \mathbf{v}_n) = \prod_{i=1}^{n} F_1(\mathbf{r}_i, \mathbf{v}_i) \tag{2-15}$$

Equations (2-15) express the fact that the probability density of finding the ith particle with position $\mathbf{r}_i$ and velocity $\mathbf{v}_i$ is independent of the positions and velocities of the other particles. This circumstance is strictly valid only if the material system is a perfect gas, since it implies that no interaction is present among the particles. In general, Eqs. (2-15) have to be modified by adding suitable correction terms to the right-hand side. The smallness of these correlation terms may allow case by case formulation of suitable closure approximations.

As an example, let us consider the case in which the interaction among the particles is small enough so that the condition

$$F_2(\mathbf{r}_1, \mathbf{r}_2, \mathbf{v}_1, \mathbf{v}_2) = F_1(\mathbf{r}_1, \mathbf{v}_1) F_1(\mathbf{r}_2, \mathbf{v}_2) \qquad (2\text{-}15)'$$

can be assumed to hold true with good approximation. Then the first equation of the hierarchy reduces to the closed form

$$\frac{\partial F_1}{\partial t}(\mathbf{r}_1, \mathbf{v}_1, t) + \mathbf{v}_1 \cdot \frac{\partial F_1}{\partial \mathbf{r}_1}(\mathbf{r}_1, \mathbf{v}_1, t) + \mathbf{K}_{11} \cdot \frac{\partial F_1}{\partial \mathbf{v}_1}(\mathbf{r}_1, \mathbf{v}_1, t)$$

$$+ (N-1) \int F_1(\mathbf{r}_2, \mathbf{v}_2, t) \mathbf{K}_{12} \cdot \frac{\partial F_1}{\partial \mathbf{v}_1}(\mathbf{r}_1, \mathbf{v}_1, t) \, d\mathbf{r}_2 \, d\mathbf{v}_2 = 0$$

$$(2\text{-}16)$$

which is known as the *Vlasov equation* (Klimontovich, 1967, Chapter III). This is the Liouville equation for a system consisting of a single particle, under the influence of an external force $m\mathbf{K}_{11}$ and the force exerted on the average by all the other particles.

The problem of solving the set of BBGKY equations remains, in general, a formidable one, even when simple closure conditions are justified. In practice, it is possible to treat with sufficient analytical detail only some cases of small deviation from the thermodynamical equilibrium distribution (*Gibbs distribution*) (Landau and Lifshitz, 1958, Chapter III):

$$F_N(\mathbf{r}_1, \ldots, \mathbf{r}_N, \mathbf{v}_1, \ldots, \mathbf{v}_N) = A \exp \frac{-H(\mathbf{r}_1, \ldots, \mathbf{r}_N, \mathbf{v}_1, \ldots, \mathbf{v}_N)}{K_B T} \qquad (2\text{-}17)$$

where H is the Hamiltonian of the system, K_B the Boltzmann constant, T the absolute temperature, and A a suitable normalization constant. It is, however, worth observing that the hierarchy of distribution functions F_n retains a fundamental conceptual importance, since it furnishes, in principle, the most complete statistical description for fluctuating systems.

II.4   Microscopic Distribution and Correlation Functions

The fundamental Eqs. (2-14) can be also deduced by means of an alternative method (Klimontovich, 1967, Chapter II) in which the microscopic density, introduced in Chapter I, is explicitly considered. We shall outline it briefly here, since it furnishes a very direct approach to the problem of evaluating ensemble averages over the scattered field. The method is based on the introduction of a hierarchy of *microscopic distribution functions* defined as

$$v_1(\mathbf{r}, \mathbf{v}, t) = \sum_{i=1}^{N} \delta[\mathbf{r} - \mathbf{r}_i(\mathbf{r}_{i0}, \mathbf{v}_{i0}, t)]\,\delta[\mathbf{v} - \mathbf{v}_i(\mathbf{r}_{i0}, \mathbf{v}_{i0}, t)] \quad (2\text{-}18)$$

$$v_2(\mathbf{r}, \mathbf{r}', \mathbf{v}, \mathbf{v}', t) = \sum_{i=1}^{N} \sum_{\substack{j=1 \\ i \ne j}}^{N} \delta[\mathbf{r} - \mathbf{r}_i(\mathbf{r}_{i0}, \mathbf{v}_{i0}, t)]\,\delta[\mathbf{v} - \mathbf{v}_i(\mathbf{r}_{i0}, \mathbf{v}_{i0}, t)]$$

$$\times\ \delta[\mathbf{r}' - \mathbf{r}_j(\mathbf{r}_{j0}, \mathbf{v}_{j0}, t)]\,\delta[\mathbf{v}' - \mathbf{v}_j(\mathbf{r}_{j0}, \mathbf{v}_{j0}, t)] \quad (2\text{-}19)$$

with obvious generalizations to higher orders, where $\mathbf{r}_i(\mathbf{r}_{i0}, \mathbf{v}_{i0}, t)$ and $\mathbf{v}_i(\mathbf{r}_{i0}, \mathbf{v}_{i0}, t)$ represent, respectively, the position and velocity at time t of the ith particle, started at $t = 0$ with the initial conditions $\mathbf{r}_{i0}, \mathbf{v}_{i0}$. These (deterministic) quantities obey a set of differential equations that can be easily obtained by taking into account the formal properties of the δ-function and the equation of motion obeyed by the ith particle. This system of equations has a form very similar to that of the BBGKY hierarchy, which can be easily deduced from it. In fact, the ensemble averages of the microscopic distribution function $v_1, v_2, \ldots, v_N$ are closely connected to the previously introduced distribution functions $F_1, F_2, \ldots, F_N$. One has precisely

$$\langle v_1(\mathbf{r}, \mathbf{v}, t) \rangle = N F_1(\mathbf{r}, \mathbf{v}, t) \quad (2\text{-}20)$$

$$\langle v_2(\mathbf{r}, \mathbf{r}', \mathbf{v}, \mathbf{v}', t) \rangle = N(N-1) F_2(\mathbf{r}, \mathbf{r}', \mathbf{v}, \mathbf{v}', t) \quad (2\text{-}21)$$

and similar relations for the higher orders, which follow directly from the definition of ensemble average:

$$\langle v_1(\mathbf{r}, \mathbf{v}, t) \rangle = \int v_1(\mathbf{r}, \mathbf{v}, t) F_N(\mathbf{r}_1, \ldots, \mathbf{r}_N, \mathbf{v}_1, \ldots, \mathbf{v}_N)$$

$$\times \, d\mathbf{r}_1 \cdots d\mathbf{r}_N \, d\mathbf{v}_1 \cdots d\mathbf{v}_N \qquad (2\text{-}22)$$

$$\langle v_2(\mathbf{r}, \mathbf{r}', \mathbf{v}, \mathbf{v}', t) \rangle = \int v_2(\mathbf{r}, \mathbf{r}', \mathbf{v}, \mathbf{v}', t) F_N(\mathbf{r}_1, \ldots, \mathbf{r}_N, \mathbf{v}_1, \ldots, \mathbf{v}_N)$$

$$\times \, d\mathbf{r}_1 \cdots d\mathbf{r}_N \, d\mathbf{v}_1 \cdots d\mathbf{v}_N \qquad (2\text{-}23)$$

once Eqs. (2-18) and (2-19) are substituted into Eqs. (2-22) and (2-23). We observe at this point that, since the expression of the microscopic scattered field does not explicitly involve the velocities of the molecules, it is convenient to introduce a class of distribution functions f_n, obtained from the F_n's by tracing over the velocities, defined by

$$f_n(\mathbf{r}_1, \ldots, \mathbf{r}_n) = \int F_n(\mathbf{r}_1, \ldots, \mathbf{r}_n, \mathbf{v}_1, \ldots, \mathbf{v}_n) \, d\mathbf{v}_1 \cdots d\mathbf{v}_n \quad (2\text{-}24)$$

The preceding relations in this way yield

$$\langle n_1(\mathbf{r}, t) \rangle = N f_1(\mathbf{r}, t) \qquad (2\text{-}25)$$

$$\langle n_2(\mathbf{r}, \mathbf{r}', t) \rangle = N(N-1) f_2(\mathbf{r}, \mathbf{r}', t) \qquad (2\text{-}26)$$

where $n_1(\mathbf{r}, t)$ is the *microscopic density*

$$n_1(\mathbf{r}, t) = \sum_{i=1}^{N} \delta[\mathbf{r} - \mathbf{r}_i(t)] \qquad (2\text{-}27)$$

and

$$n_2(\mathbf{r}, \mathbf{r}', t) = \sum_{i=1}^{N} \sum_{\substack{j=1 \\ i \neq j}}^{N} \delta[\mathbf{r} - \mathbf{r}_i(t)] \delta[\mathbf{r}' - \mathbf{r}_j(t)] \qquad (2\text{-}28)$$

The microscopic description can be completed by writing the relations connecting the *correlation functions* to the microscopic density fluctuations. These fluctuations Δn_1 are defined by

$$\Delta n_1(\mathbf{r}, t) = n_1(\mathbf{r}, t) - \langle n_1(\mathbf{r}, t) \rangle \qquad (2\text{-}29)$$

and are obviously stochastic variables with zero average. By averaging the relation

$$n_2(\mathbf{r}, \mathbf{r}', t) = n_1(\mathbf{r}, t) n_1(\mathbf{r}', t) - \delta(\mathbf{r} - \mathbf{r}') n_1(\mathbf{r}, t) \qquad (2\text{-}30)$$

which follows from Eqs. (2-27) and (2-28), we obtain, with the help of Eqs. (2-25), (2-26), and (2-29),

$$N(N-1) f_2(\mathbf{r}, \mathbf{r}', t) = N^2 f_1(\mathbf{r}, t) f_1(\mathbf{r}', t) + \langle \Delta n_1(\mathbf{r}, t) \Delta n_1(\mathbf{r}', t) \rangle$$
$$- N \delta(\mathbf{r} - \mathbf{r}') f_1(\mathbf{r}, t) \qquad (2\text{-}31)$$

If one defines the two-particle correlation function $g_2(\mathbf{r}, \mathbf{r}', t)$ as

$$g_2(\mathbf{r}, \mathbf{r}', t) = f_2(\mathbf{r}, \mathbf{r}', t) - f_1(\mathbf{r}, t) f_1(\mathbf{r}', t) \qquad (2\text{-}32)$$

Eq. (2-31) yields, in the limit of large N for which $N(N-1) \simeq N^2$,

$$N^2 g_2(\mathbf{r}, \mathbf{r}', t) = \langle \Delta n_1(\mathbf{r}, t) \Delta n_1(\mathbf{r}', t) \rangle - N \delta(\mathbf{r} - \mathbf{r}') f_1(\mathbf{r}, t) \qquad (2\text{-}33)$$

Therefore, apart from terms diverging for $\mathbf{r} = \mathbf{r}'$, and that are zero otherwise, a description of the two-particle correlation present in the system is achieved in terms of the ensemble average of the product of two density fluctuations, evaluated at different points.

The higher-order correlation functions, defined as

$$g_n(\mathbf{r}_1, \ldots, \mathbf{r}_n, t) = f_n(\mathbf{r}_1, \ldots, \mathbf{r}_n, t) - f_1(\mathbf{r}_1, t) \cdots f_1(\mathbf{r}_n, t) \qquad (2\text{-}34)$$

are no longer expressible as simple density fluctuation products of the corresponding order but are given by a linear combination of products of order $2, \ldots, n$. Thus it is more convenient to redefine a new set of correlation functions $\mathscr{G}_n$ as

$$\mathscr{G}_n(\mathbf{r}_1, \ldots, \mathbf{r}_n, t) = \frac{\langle \Delta n_1(\mathbf{r}_1, t) \cdots \Delta n_1(\mathbf{r}_n, t) \rangle}{N^n} \qquad (2\text{-}35)$$

in terms of which the g_n's can be always expressed.

For further use, it is worth generalizing the previous definition of nth order correlation function given by Eq. (2-35) by considering the case in which the density fluctuations are evaluated at different times. To this end, we generalize Eq. (2-35) as

$$\mathscr{G}_n(\mathbf{r}_1, \ldots, \mathbf{r}_n, t_1, \ldots, t_n) = \frac{\langle \Delta n_1(\mathbf{r}_1, t_1) \cdots \Delta n_1(\mathbf{r}_n, t_n) \rangle}{N^n} \quad (2\text{-}36)$$

The statistical information contained in the $\mathscr{G}_n$'s as defined in Eq. (2-36) is completely equivalent to that obtained by explicitly introducing a generalized N-particle distribution function $F_N(\{\mathbf{q}_i, \mathbf{p}_i\}, \{t_i\})$ considered at different times, representing in the Γ-space the relative density of systems whose ith particle ($i = 1, \ldots, N$) possesses coordinates $\mathbf{q}_i, \mathbf{p}_i$ at time t_i.

II.5 The Power Spectrum of a Stochastic Variable

The basic quantity usually measured in a scattering experiment in order to obtain information on the scattering medium is the distribution of the scattered intensity as a function of the scattering angle and the frequency.

The frequency distribution is a particular example of what is referred to as the *power spectrum*, a quantity that can, in general, be introduced in connection with a large class of physical situations. More precisely, let us consider a process characterized by a real function $x(t)$ whose square $x^2(t)$ possesses, for example, the physical meaning of energy. In order to describe the distribution over frequency of $x^2(t)$ it is customary to introduce the *running spectrum* of $x(t)$ (Page, 1952)

$$X(t, \omega) = \frac{1}{2\pi} \int_{-\infty}^{t} x(\tau) e^{i\omega\tau} d\tau \quad (2\text{-}37)$$

The energy of $x(t)$ up to time t, according to Eq. (2-37), is given by

$$\int_{-\infty}^{t} x^2(\tau) d\tau = 2\pi \int_{-\infty}^{+\infty} |X(t, \omega)|^2 d\omega \quad (2\text{-}38)$$

so that $2\pi |X(t, \omega)|^2 d\omega$ can be interpreted as the energy contained up to time t in the frequency interval $d\omega$ centered around ω. Therefore, the instantaneous power spectrum is given by

$$Y(t,\omega) = 2\pi \frac{\partial}{\partial t} |X(t,\omega)|^2 \qquad (2\text{-}39)$$

From Eqs. (2-37) and (2-39) it follows directly that

$$Y(t,\omega) = \frac{1}{\pi} \int_0^{+\infty} x(t)x(t+\tau)\cos\omega\tau \, d\tau \qquad (2\text{-}40)$$

Whenever $x(t)$ represents a stochastic process Eq. (2-40) has a simple generalization in terms of ensemble averages, so that

$$Y(t,\omega) = \frac{1}{\pi} \int_0^{+\infty} \langle x(t)x(t+\tau)\rangle \cos\omega\tau \, d\tau \qquad (2\text{-}41)$$

and

$$Y(t,\omega) = 2\pi \frac{\partial}{\partial t} \langle |X(t,\omega)|^2 \rangle \qquad (2\text{-}42)$$

These considerations hold, in general, for nonstationary and stationary processes. In particular, Eq. (2-41) specializes, in the latter case, to give the well-known *Wiener–Khintchine* theorem (Born and Wolf, 1970; Chapter X). One can also rewrite Eq. (2-41) in terms of the analytic signal $\hat{x}(t)$ as [see Eq. (1-60)]

$$Y(t,\omega) = \frac{1}{4\pi} \int_0^{+\infty} \langle [\hat{x}(t)+\hat{x}^*(t)][\hat{x}(t+\tau)+\hat{x}^*(t+\tau)]\rangle$$
$$\times \cos\omega\tau \, d\tau \qquad (2\text{-}43)$$

Under the hypothesis of stationarity, averages of the kind $\langle \hat{x}(t)\hat{x}(t+\tau)\rangle$ are zero, while the autocorrelation function $\langle \hat{x}^*(t)\hat{x}(t+\tau)\rangle$ depends only on the difference $(t+\tau)-t = \tau$, so that

$$Y(\omega) = \frac{1}{4\pi} \int_0^{+\infty} [\langle \hat{x}^*(t)\hat{x}(t+\tau)\rangle + \langle \hat{x}(t)\hat{x}^*(t+\tau)\rangle]\cos\omega\tau \, d\tau$$
$$(2\text{-}44)$$

which, due to the fact that the integrand is an even function of τ, yields

$$Y(\omega) = \frac{1}{8\pi} \int_{-\infty}^{+\infty} \left[\langle \hat{x}^*(t)\,\hat{x}(t+\tau) \rangle + \langle \hat{x}(t)\,\hat{x}^*(t+\tau) \rangle \right] e^{i\omega\tau} \, d\tau$$
(2-45)

Since $Y(\omega)$ is an even function of ω, all physical information is contained in the values assumed by $Y(\omega)$ for $\omega \geqslant 0$. If one then considers only positive frequencies, Eq. (2-45) gives

$$Y(\omega) = \frac{1}{8\pi} \int_{-\infty}^{+\infty} \langle \hat{x}(t+\tau)\,\hat{x}^*(t) \rangle e^{i\omega\tau} \, d\tau \qquad \text{for} \quad \omega \geqslant 0 \quad (2\text{-}46)$$

due to the fact that no contribution is given by the term $\langle \hat{x}(t)\,\hat{x}^*(t+\tau) \rangle$, as can easily be seen by using Eqs. (1-59) and (2-45). For practical purposes, it is useful to define the *power spectrum* $I(\omega)$ as

$$I(\omega) = \frac{1}{8\pi} \int_{-\infty}^{+\infty} \langle \hat{x}(t+\tau)\,\hat{x}^*(t) \rangle \, e^{i\omega\tau} \, d\tau$$
(2-47)

for all values of ω, so that $I(\omega) = Y(\omega)$ for $\omega \geqslant 0$ and $I(\omega) = 0$ for $\omega < 0$ [see Eq. (1-59)].

II.6 The Spectrum of the Scattered Light

Let us apply the preceding considerations to the case of molecular scattering in which the stochastic quantity is represented by the scattered electric field in the single-scattering approximation, as given in Eq. (1-89). It is convenient to modify Eq. (2-47) into

$$I(\omega) = \frac{c}{16\pi^2} \int_{-\infty}^{+\infty} \langle \hat{\mathbf{E}}(t+\tau) \cdot \hat{\mathbf{E}}^*(t) \rangle e^{i\omega\tau} \, d\tau \qquad (2\text{-}48)$$

since this yields, with the help of Eq. (1-60),

$$\int_0^{+\infty} I(\omega) \, d\omega = \frac{c}{8\pi} \langle \hat{\mathbf{E}}(t) \cdot \hat{\mathbf{E}}^*(t) \rangle = \frac{c}{4\pi} \langle \mathbf{E}(t) \cdot \mathbf{E}(t) \rangle \quad (2\text{-}49)$$

which allows us to interpret the quantity on the left-hand side of Eq. (2-49) as the electromagnetic intensity.

If one now introduces Eq. (1-89) into Eq. (2-48), one obtains for the spectrum of the scattered intensity in the direction $\boldsymbol{\eta} = \mathbf{r}/r$:

$$I(\omega, \boldsymbol{\eta})$$

$$= \mathbf{Z}^2 \int_{-\infty}^{+\infty} d\tau \exp[i(\omega - \omega_0)\tau]$$

$$\times \left\langle \sum_{i,j}^{1,} \exp\left[i\mathbf{k}_1 \cdot \mathbf{R}_j\left(t + \tau - \frac{r}{c}\right)\right] \exp\left[-i\mathbf{k}_1 \cdot \mathbf{R}_i\left(t - \frac{r}{c}\right)\right]\right\rangle$$

$$\tag{2-50}$$

where

$$\mathbf{Z} = -\frac{\sqrt{c}\,\alpha(\omega_0)\,k_0^2}{r}\,\boldsymbol{\eta} \times (\boldsymbol{\eta} \times \mathbf{E}_0) \tag{2-51}$$

The definition of microscopic density n_1 given by Eq. (2-27) allows us to write

$$\sum_{i=1}^{N} \exp[i\mathbf{k}_1 \cdot \mathbf{R}_i(t)] = \sum_{i=1}^{N} \int_{V_{sc}} \delta[\mathbf{r}' - \mathbf{R}_i(t)] \exp(i\mathbf{k}_1 \cdot \mathbf{r}')\, d\mathbf{r}'$$

$$= \int_{V_{sc}} \exp(i\mathbf{k}_1 \cdot \mathbf{r}') n_1(\mathbf{r}', t)\, d\mathbf{r}' \tag{2-52}$$

where V_{sc} is the scattering volume. Inserting Eq. (2-52) into Eq. (2-50) gives

$$I(\omega, \boldsymbol{\eta}) = \mathbf{Z}^2 \int_{-\infty}^{+\infty} \exp[i(\omega - \omega_0)\tau]\, d\tau \int_{V_{sc}} d\mathbf{r}' \int_{V_{sc}} d\mathbf{r}''$$

$$\times \exp[i\mathbf{k}_1 \cdot (\mathbf{r}' - \mathbf{r}'')]\left\langle n_1\left(\mathbf{r}', t + \tau - \frac{r}{c}\right) n_1\left(\mathbf{r}'', t - \frac{r}{c}\right)\right\rangle$$

$$\tag{2-53}$$

This is the formula obtained by Komarov and Fisher (1962) as an extension of the analogous relations first derived by

Glauber (1952, 1954) and van Hove (1954) in the case of neutron scattering by crystals and liquids.

The spectrum $I(\omega, \eta)$ of scattered light can be conveniently expressed in terms of density fluctuations with the help of Eq. (2-29) as

$$
I(\omega, \eta) = \mathbf{Z}^2 \int_{-\infty}^{+\infty} \exp\left[i(\omega - \omega_0)\tau\right] \int_{V_{sc}} d\mathbf{r}' \int_{V_{sc}} d\mathbf{r}''
$$

$$
\times \exp\left[i\mathbf{k}_1 \cdot (\mathbf{r}' - \mathbf{r}'')\right]\left[\left\langle n_1\left(\mathbf{r}', t + \tau - \frac{r}{c}\right)\right\rangle\right.
$$

$$
\times \left\langle n_1\left(\mathbf{r}'', t - \frac{r}{c}\right)\right\rangle + \left\langle \Delta n_1\left(\mathbf{r}', t + \tau - \frac{r}{c}\right)\right.
$$

$$
\left.\left. \times \Delta n_1\left(\mathbf{r}'', t - \frac{r}{c}\right)\right\rangle\right] \tag{2-54}
$$

which can be rewritten, with the help of Eqs. (2-25) and (2-36), in the form

$$
I(\omega, \eta) = \mathbf{Z}^2 N^2 \int_{-\infty}^{+\infty} \exp\left[i(\omega - \omega_0)\tau\right] d\tau \int_{V_{sc}} d\mathbf{r}' \int_{V_{sc}} d\mathbf{r}''
$$

$$
\times \exp\left[i\mathbf{k}_1 \cdot (\mathbf{r}' - \mathbf{r}'')\right]\left[f_1\left(\mathbf{r}', t + \tau - \frac{r}{c}\right) f_1\left(\mathbf{r}'', t - \frac{r}{c}\right)\right.
$$

$$
\left. + \mathscr{G}_2\left(\mathbf{r}', \mathbf{r}'', t + \tau - \frac{r}{c}, t - \frac{r}{c}\right)\right] \tag{2-55}
$$

In particular, if the properties of the medium are invariant under spatial translations, that is, for homogeneous systems, f_1 does not depend on space, while $\mathscr{G}_2$ depends on $\mathbf{r}'$ and $\mathbf{r}''$ through the difference $\rho = \mathbf{r}' - \mathbf{r}''$. The same property is valid in the time domain for stationary systems, which amounts to saying that the system is invariant under temporal translations. The validity of this condition has already been implicitly assumed since it ensures the stationarity of the scattered field thus justifying the use of Eq. (2-48).

The stationarity and homogeneity conditions permit us to

simplify Eq. (2-55). The term containing the product $f_1 f_1$ can be neglected for all scattering geometries such that $V_{sc}^{1/3} k_1 \gg 1$, due to the presence of the oscillating factor $\exp(i\mathbf{k}_1 \cdot \mathbf{r})$. This condition, which specializes in terms of the scattering angle ϑ (see Fig. 1.1) as $2k_0 \sin(\vartheta/2) V_{sc}^{1/3} \gg 1$, is actually always met in practice at visible wavelengths for ϑ considerably different from zero. Under this circumstance, Eq. (2-55) reduces to

$$I(\omega, \mathbf{\eta}) = \mathbf{Z}^2 N^2 V_{sc} \int_{-\infty}^{+\infty} d\tau \int d\mathbf{\rho} \exp\left[i\mathbf{k}_1 \cdot \mathbf{\rho} + i(\omega - \omega_0)\tau\right] \mathscr{G}_2(\mathbf{\rho}, \tau) \tag{2-56}$$

which has been obtained using the hypothesis that the typical correlation length l_c of the medium, that is, the length for which

$$\mathscr{G}_2(\mathbf{\rho}, \tau) \simeq 0 \qquad \text{for} \quad \rho > l_c \tag{2-57}$$

fulfills the relation

$$\frac{V_{sc}^{1/3}}{l_c} \gg 1 \tag{2-58}$$

Equation (2-56) amounts to saying that the spectrum of scattered light in a direction corresponding to a transferred momentum $\mathbf{k}_1$ is proportional to the Fourier space–time transform

$$\tilde{\mathscr{G}}_2(\mathbf{k}, \omega) = \frac{1}{2\pi} \int_{-\infty}^{+\infty} d\tau \int d\mathbf{\rho} \exp\left[-i\mathbf{k} \cdot \mathbf{\rho} + i\omega\tau\right] \mathscr{G}_2(\mathbf{\rho}, \tau) \tag{2-59}$$

of the second-order correlation function, evaluated at the wave-vector $-\mathbf{k}_1$ and frequency $\omega - \omega_0$.

The relation

$$k_1 = 2k_0 \sin\left(\frac{\vartheta}{2}\right) \leqslant 2k_0$$

implies that $\tilde{\mathscr{G}}_2(-\mathbf{k}_1, \omega - \omega_0)$ cannot be measured for $k_1 > 2k_0$, so that a complete determination of the spatial behavior of $\mathscr{G}_2(\mathbf{\rho}, \tau)$ is possible only if $l_c > 1/2k_0$. In this case $\mathscr{G}_2(\mathbf{\rho}, \tau)$ can be obtained by simply inverting Eq. (2-56) and taking into account the fact that $\tilde{\mathscr{G}}_2(\mathbf{k}, \omega - \omega_0) \simeq 0$ for $k > 2k_0$. In practice, however, this condition is never met in light-scattering experiments performed on molecular systems under ordinary conditions, for which k_0 is typically of the order of $10^5 \, \mathrm{cm}^{-1}$ and l_c of the order of a few angstroms. This is equivalent to saying that

$$\tilde{\mathscr{G}}_2(2\mathbf{k}_0, \omega - \omega_0) \simeq \tilde{\mathscr{G}}_2(0, \omega - \omega_0)$$

$$= \frac{1}{2\pi} \int_{-\infty}^{+\infty} d\tau \int d\mathbf{\rho} \exp\left[i(\omega - \omega_0)\tau\right] \mathscr{G}_2(\mathbf{\rho}, \tau)$$

$$(2\text{-}60)$$

so that only integral information on the spatial behavior of $\mathscr{G}_2(\mathbf{\rho}, \tau)$ is available.

Conversely, no limitations exist, in principle, on the determination of the temporal behavior of $\mathscr{G}_2$, so that the measurement of the frequency spectrum of scattered light furnishes a complete description, in the time domain, of the second-order statistical properties of the medium. It will be seen later what kind of measurements on scattered light have to be performed in order to obtain more general information concerning higher-order correlation functions.

II.7 Further Sources of Fluctuation for Scattered Light

The fluctuating character of the scattered field is evident from Eqs. (1-89) and (2-56), which clearly show the influence of the stochastic molecular motion on the scattered radiation. It is evident that the statistical nature of the scattered field is not only due to the translational motion of molecules. As a matter of fact, the derivation of Eq. (2-56) from Eq. (1-89) is based on the assumption that no temporal fluctuation is undergone by the incident field. Furthermore, another source of fluctuation is present when dealing with optically anisotropic molecules, whose polarizability is a tensor.

The first circumstance occurs whenever the incident radiation is well represented by an ideal monochromatic plane polarized beam (as, for example, the one obtained by means of a laser with highly stabilized amplitude) but is far from being verified for other kinds of radiation (as, for example, the one furnished by a natural source). More precisely, the electromagnetic perturbation produced by many kinds of sources possesses an intrinsically chaotic character, which is also retained after obtaining monochromatic light by means of a suitable filtering operation. One can deal with this situation by generalizing the ensemble average operation to include the statistical variable represented by the incident field (Mandel, 1969; Bertolotti *et al.*, 1970).

The assumption of time independence of the polarizability tensor α is obviously verified for optically symmetrical molecules, that is, when α reduces to a scalar quantity. This is no longer true for nonspherical molecules, since their rotational motion induces temporal fluctuations of α, which in turn implies the necessity of suitably modifying Eq. (1-89). Generalization of the scattering formalism has been achieved (Steele and Pecora, 1965) by introducing, besides the positions of the center of mass of the molecule, a new set of dynamical variables $\Omega_i \equiv (\Omega_{1i}, \Omega_{2i}, \Omega_{3i})$, that is, the Eulerian angles of orientation of the ith molecule with respect to a fixed system of axes. This leads in a natural way to the introduction of a distribution function $W[\mathbf{R}_1(t), \Omega_1(t), \mathbf{R}_2(0), \Omega_2(0)]$, to be used when evaluating ensemble averages, representing the probability of finding molecule 1 with center-of-mass position $\mathbf{R}_1$ and orientation Ω_1 at time t, if molecule 2 is known to have position $\mathbf{R}_2(0)$ and orientation $\Omega_2(0)$ at time zero. The spectrum of the scattered light can then be expressed in terms of this generalized distribution function.

The tensor character of the polarizability is also a source of depolarization of the scattered radiation. In fact, for spherically symmetrical molecules the induced dipole moment is always parallel to the incident field, irrespective of their orientations, while this is not true for nonspherical molecules, whose rotations cause the direction of the induced dipoles to fluctuate (see, for example, Gershon and Oppenheim, 1973).

In order to treat quantitatively the depolarization effect, it is customary to introduce a suitable quantity called the *depolariz- ation factor* Δ (Fabelinskii, 1968, Chapter I). More precisely, one considers the radiation scattered at $90°$, in the direction orthogonal to $\mathbf{E}_0$ and $\mathbf{k}_0$. If α is a scalar, then Eq. (1-89) ensures that the scattered electric field has the same direction as $\mathbf{E}_0$, while the tensorial nature of α gives rise to a nonvanishing component along the direction of $\mathbf{k}_0$. The ratio between the intensities of light scattered with electric field parallel to $\mathbf{k}_0$ and $\mathbf{E}_0$ is precisely Δ.[†] The measurement of this quantity enables us to convey information on the optical structure of the scattering molecules.

It is worth observing that the statement that light scattered by isotropically polarizable molecular fluids composed of spheri- cally symmetric molecules is polarized, holds true under the assumption of the validity of Eq. (1-89), that is, whenever the higher-order scattering terms appearing in Eq. (1-84) can be considered negligible. There is, however, experimental evidence that light scattered by such fluids possesses a nonnegligible depolarized component, also if its intensity is, under ordinary conditions, thousands of times smaller than that pertaining to the polarized component (see, for an extensive review, the work of Gelbart, 1974).

Mazur (1958) and Tanaka (1968) have provided an inter- pretation of these effects on the basis of the statistical mechanics of the electromagnetic properties of many-body systems. An exhaustive discussion on depolarized light scattering by simple fluids has been recently furnished by Gelbart (1974), who is able to interpret in the frame of a general microscopic

[†]The choice of $90°$ angle derives from the fact that, for natural light and scalar α, the electric field scattered at $90°$ always has a definite direction, being orthogonal to $\mathbf{k}_0$ and to the scattering direction, so that the definition of depolarization factor can be extended to this kind of radiation.

treatment previous approaches accounting for depolarized scattering by gases (Levine and Birnbaum, 1968) and liquids (Bucaro and Litovitz, 1971) on the basis of a binary collision model.

Macroscopic Approach to Scattering Theory

III.1 Microscopic and Macroscopic Points of View

In the first two chapters, we outlined the basis of a microscopic molecular scattering theory whose validity implies the assumption of two hypotheses. First, the frequency broadening $\Delta\omega$ of the scattered light must verify the relation $\Delta\omega/\omega_0 \ll 1$; second, no relevant change in the internal molecular state must occur due to the incident radiation. The first condition is essentially dictated by the necessity of considering the polarizability $\alpha(\omega)$ constant over the frequency range relative to the spectrum of the scattered field [see Eq. (1-53)], and it is practically equivalent to the nonrelativistic hypothesis $v/c \ll 1$. The second condition corresponds to neglecting the so-called *combination scattering*, or *Raman scattering*, which is derived from changes in the internal structure of molecules. This amounts to saying that we limit ourselves to the consideration of *Rayleigh scattering*, which corresponds to the scattering of light connected with the motion of each molecule considered as a whole, that is, without taking into account its internal dynamics. Rayleigh scattering can be studied, in principle, with any degree of accuracy by resorting to the microscopic fundamental approach. This procedure is by definition exact and requires a detailed knowledge of the dynamics of the system.

One can ask at this point if it is possible to resort to the macroscopic approach, that is, to deal with quantities expressible as averages over a great number of scatterers, which are expected to be more useful than the relevant microscopic quantities. This procedure, which is the basic one of the electrodynamics of continuous media, consists of introducing suitable variables characterizing the overall electromagnetic behavior of the material system. As is well known, for a nonconducting, non-magnetic, nonabsorbing medium one can simply look for the existence of the *dielectric susceptibility* χ relating the electric field $\mathbf{E}$ and the electric polarization $\mathbf{P}$ as

$$\mathbf{P}(\mathbf{r}, t) = \chi \mathbf{E}(\mathbf{r}, t) \tag{3-1}$$

Here both $\mathbf{E}$ and $\mathbf{P}$ are macroscopic quantities, $\mathbf{E}$ being the ensemble average over the possible realizations of the fluctuating microscopic electric field evaluated at $\mathbf{r}$ and t, and $\mathbf{P}$ representing the dipole moment per unit volume, that is,

$$\mathbf{P}(\mathbf{r}, t) = \left\langle \sum_i \mathbf{P}_i(t) \delta\left[\mathbf{r} - \mathbf{R}_i(t)\right] \right\rangle \tag{3-2}$$

if $\mathbf{P}_i(t)$ is the dipole moment and $\mathbf{R}_i(t)$ the trajectory of the *i*th molecule. Whenever Eq. (3-1) applies, it is possible to write the macroscopic Maxwell equations (Born and Wolf, 1970, Chapter II):

$$\nabla \times \mathbf{E} = -\frac{1}{c}\frac{\partial \mathbf{H}}{\partial t}, \qquad \nabla \cdot (\varepsilon_0 \mathbf{E}) = 0$$

$$\nabla \times \mathbf{H} = \frac{1}{c}\frac{\partial}{\partial t}(\varepsilon_0 \mathbf{E}), \qquad \nabla \cdot \mathbf{H} = 0 \tag{3-3}$$

where ε_0 is the *dielectric constant* defined by the relation

$$\varepsilon_0 = 1 + 4\pi\chi \qquad (3\text{-}4)$$

It can immediately be seen that this approach is not sufficient to describe the problem of scattering, independently of the existence of the dielectric constant. As a matter of fact, ε_0 is, for example, independent of $\mathbf{r}$ for a homogeneous medium, which in turn implies that the solution of the macroscopic Maxwell equations does not correspond to any scattered field. However, it is possible in some cases to modify the definition of the dielectric constant in such a way as to obtain a meaningful macroscopic treatment of scattering. As a matter of fact the first treatment of light scattering by a real fluid, due to Einstein (1910) and Smolouchosky (1908), evolved precisely along this line. Their main result is contained in the expression of the *extinction coefficient h*, which is defined as the ratio of the total intensity of light scattered in all directions per unit volume of the diffusing medium to the incident flux density (see Landau and Lifshitz, 1960, Chapter XIV). The quantity h was experimentally accessible before the introduction of the modern sophisticated spatial and temporal resolution techniques. From the historical point of view, h was first derived on a microscopic basis by Rayleigh (1881), who found, for the simple case of light scattered by a gas (independent scatterers),

$$h = \frac{8\pi}{3}\frac{\omega_0^4}{c^4}\alpha^2\rho_0 \qquad (3\text{-}5)$$

where ω_0 is the angular frequency of the incident light, α the molecular polarizability at this frequency, and ρ_0 the average number density.

The Einstein–Smolouchosky formula, derived on a macroscopic basis, is

$$h = \frac{\omega_0^4}{6\pi c^4}\left[K_B T\rho_0^2\chi_T\left(\frac{\partial\varepsilon_0}{\partial\rho}\right)^2_{\rho=\rho_0} + \frac{K_B T^2}{\rho_0 c_v}\left(\frac{\partial\varepsilon_0}{\partial T}\right)^2_{\rho=\rho_0} \right] \qquad (3\text{-}6)$$

where $\chi_T = (1/\rho_0)(\partial\rho/\partial p)_{T,\,\rho=\rho_0}$ is the isothermal compressibility, K_B Boltzmann's constant, and T, ρ_0, p, and c_v represent, respectively, equilibrium temperature, number density, pressure, and specific heat per unit mass at constant volume, the dielectric constant ε_0 having been characterized thermodynamically by the two parameters ρ and T. We observe that ε_0, as defined in terms of ensemble average, is connected to ρ_0 and T by means of a suitable equation, a celebrated version of which is the so-called *Lorenz–Lorentz* (or *Clausius–Mossotti* in the zero-frequency limit) equation (Born and Wolf, 1970, Chapter II):

$$\frac{\varepsilon_0 - 1}{\varepsilon_0 + 2} = \frac{4\pi}{3}\alpha\rho_0 \qquad (3\text{-}7)$$

We present in the next section a macroscopic treatment of scattering that also takes into account the frequency broadening of the incident radiation, and that, in particular, allows one to recover Eq. (3-6). This is done by postulating the existence, which is not a *priori* assured by that of ε_0, of a fluctuating phenomenological dielectric constant $\varepsilon(\mathbf{r}, t)$, by means of which one writes a set of equations formally analogous to Eq. (3-3), which can yield the expression of the single realization of the scattered electric field instead of its vanishing ensemble average. The ultimate justification for this procedure will be discussed by comparing the expression of the extinction coefficient h, as given by Eq. (3-6), with that furnished by the rigorous microscopic theory. However, this can yield only a necessary condition for the correctness of the frequency and angular dependence of the spectrum.

III.2   The Spectrum of Light Scattered by Density Fluctuations

We assume the validity of the macroscopic Maxwell equations given by Eqs. (3-3) with a fluctuating dielectric constant $\varepsilon(\mathbf{r}, t)$. In some sense this amounts to defining instantaneous fluctuating quantities by means of space averages over a physical infinitesimal volume v, that is, a volume small with respect to the wavelength λ_0 but still large enough to contain a great number of molecules, without averaging the result with respect to the motion of the particles (Landau and Lifshitz 1960, Chapter XIV).[†] As an example, the fluctuating density is defined as $\rho(\mathbf{r}, t) = N(t)/v$, where $N(t)$ is the number of particles contained in the volume v around $\mathbf{r}$ at time t.

The solution of the set of Eqs. (3-3) can be obtained by means of a perturbative approach (Pecora, 1964; van Kampen, 1969), which is valid if the total scattered intensity is small compared to that of the incident radiation. More precisely, let us suppose ε is the sum of its constant ensemble average value ε_0 plus a small fluctuating part ε_1, as

$$\varepsilon(\mathbf{r}, t) = \varepsilon_0 + \varepsilon_1(\mathbf{r}, t) \tag{3-8}$$

(We assume in the following that ε_0 is numerically equal to unity for the sake of simplicity.) Correspondingly, let us write the electric field as

$$\hat{\mathbf{E}}(\mathbf{r}, t) = \hat{\mathbf{E}}_{\text{ext}}(\mathbf{r}, t) + \hat{\mathbf{E}}_1(\mathbf{r}, t) \tag{3-9}$$

with

$$\hat{\mathbf{E}}_{\text{ext}}(\mathbf{r}, t) = \mathbf{E}_0 \exp(i\mathbf{k}_0 \cdot \mathbf{r} - i\omega_0 t) \tag{3-10}$$

where $k_0 = \omega_0/c$, and $\mathbf{k}_0 \cdot \mathbf{E}_0 = 0$ (and analogous relations for the magnetic field). Henceforth we substitute for the electric and magnetic fields their analytic representations. It is easy to deduce the associated Maxwell equations from the corresponding

---

[†] The characteristic time scale T_1 of these fluctuations has to be compared with the time scale T_2 associated with the frequency dependence of the single-molecule polarizability. Only if $T_1 \gg T_2$, as is often the case at optical frequencies, can the relation between $\mathbf{P}(t)$ and $\mathbf{E}(t)$ have the simple form given by Eq. (3-3), with χ fluctuating over the time T_1. In the general case, $\mathbf{P}(t)$ is expressed by means of a convolution integral of χ and $\mathbf{E}$, and the spectral properties of the scattered field are modified accordingly (Mandel and Wolf, 1973).

equations for the real quantities, under the condition of a slowly varying ε by means of a procedure analogous to the one used in Section I.4. The set of Eqs. (3-3) can then be linearized to furnish a zeroth-order solution given by $\hat{\mathbf{E}}_{\text{ext}}$ and $\hat{\mathbf{H}}_{\text{ext}}$, while the first-order equations are

$$\nabla \times \hat{\mathbf{E}}_1 = -\frac{1}{c}\frac{\partial \hat{\mathbf{H}}_1}{\partial t}, \qquad\qquad \nabla \cdot \hat{\mathbf{E}}_1 = -\nabla \cdot (\varepsilon_1 \hat{\mathbf{E}}_{\text{ext}})$$

$$\nabla \times \hat{\mathbf{H}}_1 = \frac{1}{c}\frac{\partial \hat{\mathbf{E}}_1}{\partial t} + \frac{1}{c}\frac{\partial}{\partial t}(\varepsilon_1 \hat{\mathbf{E}}_{\text{ext}}), \qquad \nabla \cdot \hat{\mathbf{H}}_1 = 0 \tag{3-11}$$

The set of Eqs. (3-11) can be rewritten in a form identical to Eqs. (1-7)–(1-10), by means of the introduction of the vector $\mathbf{p}_M$ defined as

$$\mathbf{p}_M(\mathbf{r}, t) = \frac{1}{4\pi}\varepsilon_1 \hat{\mathbf{E}}_{\text{ext}} = \frac{1}{4\pi}\varepsilon_1(\mathbf{r}, t)\,\mathbf{E}_0 \exp(i\mathbf{k}_0 \cdot \mathbf{r} - i\omega_0 t) \tag{3-12}$$

Thus the same procedure followed in Section I.1 in order to derive the Hertz vector $\mathbf{\Pi}$ furnishes the macroscopic Hertz vector $\hat{\mathbf{\Pi}}_1$ as [see Eq. (1-14)]

$$\hat{\mathbf{\Pi}}_1(\mathbf{r}, t) = \int \frac{\mathbf{p}_M\left[\mathbf{r}', t - (|\mathbf{r} - \mathbf{r}'|/c)\right]}{|\mathbf{r} - \mathbf{r}'|}\, d\mathbf{r}' \tag{3-13}$$

The usual assumption $r \gg V_{\text{sc}}^{1/3}$ (V_{sc} being the scattering volume) gives $|\mathbf{r} - \mathbf{r}'| \simeq r - \mathbf{r}' \cdot (\mathbf{r}/r)$, while the denominator in the integrand of Eq. (3-13) can be approximated by r. In this way, we can write

$$\hat{\mathbf{\Pi}}_1(\mathbf{r}, t) = \frac{1}{r}\int \mathbf{p}_M\left(\mathbf{r}', t - \frac{r}{c} + \frac{\mathbf{r} \cdot \mathbf{r}'}{cr}\right) d\mathbf{r}'$$

$$= \frac{\mathbf{E}_0}{4\pi r}\int \exp\left[i\mathbf{k}_1 \cdot \mathbf{r}' - i\omega_0\left(t - \frac{r}{c}\right)\right]$$

$$\times \varepsilon_1\left(\mathbf{r}', t - \frac{r}{c} + \frac{\mathbf{r} \cdot \mathbf{r}'}{cr}\right) d\mathbf{r}' \tag{3-14}$$

where $\mathbf{k}_1 = \mathbf{k}_0 - (\omega_0/c)\boldsymbol{\eta}$ (with $\boldsymbol{\eta} = \mathbf{r}/r$). The expression of the electric field $\hat{\mathbf{E}}_1$ can then be derived by means of Eqs. (1-11). The assumption $|\partial \varepsilon_1/\partial t| \ll \omega_0|\varepsilon_1|$ allows us to write, with good approximation,

$$\frac{\partial}{\partial t} = -i\omega_0 \tag{3-15}$$

while neglecting the terms proportional to $1/r^2$ and $1/r^3$ with respect to those proportional to $1/r$ yields

$$\nabla\left(\nabla \cdot \mathbf{E}_0 \frac{e^{ik_0r}}{r}\right) = -\frac{e^{ik_0r}}{r} k_0^2 \boldsymbol{\eta}(\boldsymbol{\eta} \cdot \mathbf{E}_0)$$

$$= [-k_0^2 \boldsymbol{\eta} \times (\boldsymbol{\eta} \times \mathbf{E}_0) - k_0^2 \mathbf{E}_0] \frac{e^{ik_0r}}{r} \tag{3-16}$$

We now observe that Eq. (3-15) and the condition $r \gg V_{sc}^{1/3}$ justifies neglecting the dependence on $\mathbf{r}$ of the factors ε_1 and $e^{i\mathbf{k}_1 \cdot \mathbf{r}}$, respectively, appearing in Eq. (3-14), so that the electric field $\hat{\mathbf{E}}_1$ takes the form

$$\hat{\mathbf{E}}_1(\mathbf{r}, t) = -k_0^2 \boldsymbol{\eta} \times [\boldsymbol{\eta} \times \hat{\boldsymbol{\Pi}}_1(\mathbf{r}, t)] \tag{3-17}$$

In order to obtain the spectrum $I(\omega, \boldsymbol{\eta})$ of the scattered intensity in the $\boldsymbol{\eta}$-direction, we recall Eq. (2-48), which, with the help of Eqs. (3-14) and (3-17) and under the hypothesis of stationarity, yields

$$I(\omega, \boldsymbol{\eta}) = \frac{c}{16\pi^2} \int_{-\infty}^{+\infty} \langle \hat{\mathbf{E}}_1(\tau) \cdot \hat{\mathbf{E}}_1^*(0) \rangle e^{i\omega\tau} \, d\tau$$

$$= \frac{k_0^4 c}{16\pi^2} \frac{E_0^2 \sin^2 \gamma}{(4\pi r)^2} \int d\mathbf{r}' \, d\mathbf{r}'' \, d\tau \exp[i(\omega - \omega_0)\tau]$$

$$\times \exp[i\mathbf{k}_1(\mathbf{r}' - \mathbf{r}'')] \left\langle \varepsilon_1\left[\mathbf{r}', \tau + \frac{1}{c}\boldsymbol{\eta} \cdot (\mathbf{r}' - \mathbf{r}'')\right] \varepsilon_1(\mathbf{r}'', 0) \right\rangle \tag{3-18}$$

where γ is the angle between $\mathbf{E}_0$ and $\mathbf{\eta}$ and we take advantage of the fact that the ensemble average $\langle \varepsilon(\mathbf{r}', t') \varepsilon(\mathbf{r}'', t'') \rangle$ is a function of the difference $t' - t''$. For a homogeneous system, that is, if the same condition also applies to the spatial coordinates, and for correlation lengths much smaller than V_{sc} [see Eqs. (2-57) and (2-58)], one has

$$I(\omega, \mathbf{\eta}) = \frac{k_0^4 c E_0^2 \sin^2 \gamma \, V_{sc}}{16\pi^2 (4\pi r)^2} \int d\mathbf{r}' \, d\tau \exp[i(\omega - \omega_0)\tau] \exp(i\mathbf{k}_1 \cdot \mathbf{r}')$$

$$\times \left\langle \varepsilon_1 \left(\mathbf{r}', \tau + \frac{1}{c} \mathbf{\eta} \cdot \mathbf{r}' \right) \varepsilon_1(0, 0) \right\rangle \tag{3-19}$$

In particular, if the term $(1/c)\,\mathbf{\eta} \cdot \mathbf{r}'$ is, for all significant values of $\mathbf{r}'$, negligible with respect to the characteristic correlation time of $\langle \varepsilon_1(\mathbf{r}', \tau) \varepsilon_1(0, 0) \rangle$, it can be dropped in the temporal argument of ε_1.

The thermodynamic behavior of ε_1 can be, for example, described by means of the two independent state variables density ρ and temperature T. If one neglects the dependence on T (which is a very good approximation for a gas at optical frequencies and for many substances in a large range of temperature), one can approximate and write

$$\varepsilon_1(\mathbf{r}, t) = \left(\frac{d\varepsilon}{d\rho} \right)_{\rho = \rho_0} [\rho(\mathbf{r}, t) - \rho_0] = \left(\frac{d\varepsilon}{d\rho} \right)_{\rho = \rho_0} \rho_1(\mathbf{r}, t) \tag{3-20}$$

where ρ_1 represents the instantaneous density fluctuation. In this way Eq. (3-19) yields

$$I(\omega, \mathbf{\eta}) = \frac{V_{sc} I_0}{(4\pi r)^2} k_0^4 \sin^2 \gamma \left(\frac{d\varepsilon}{d\rho} \right)^2 S(-\mathbf{k}_1, \omega - \omega_0) \tag{3-21}$$

where the dynamic form factor $S(\mathbf{k}, \omega)$ is the double Fourier transform of the *correlation function* of the density fluctuations

$$S(\mathbf{k}, \omega) = \frac{1}{2\pi} \int \langle \rho_1(\mathbf{r}', t') \rho_1(0,0) \rangle \exp(-i\mathbf{k}\cdot\mathbf{r}' + i\omega t') \, d\mathbf{r}' \, dt' \quad (3\text{-}22)$$

and $I_0 = (c/8\pi) E_0^2$ represents the incident intensity.

The total scattered intensity in a given direction $\boldsymbol{\eta}$ can be evaluated with the help of Eqs. (3-21) and (3-22) by integrating over the frequency domain, thus obtaining

$$I(\boldsymbol{\eta}) = \int_{-\infty}^{+\infty} d\omega \, I(\omega, \boldsymbol{\eta}) = \frac{V_{sc} I_0}{(4\pi r)^2} k_0^4 \sin^2 \gamma \left(\frac{d\varepsilon}{d\rho}\right)^2 S(-\mathbf{k}_1) \quad (3\text{-}23)$$

where

$$S(-\mathbf{k}_1) = \int \langle \rho_1(\mathbf{r}', 0) \rho_1(0,0) \rangle e^{i\mathbf{k}_1\cdot\mathbf{r}'} \, d\mathbf{r}' \quad (3\text{-}24)$$

having taken advantage of the δ-function property

$$\delta(x) = \frac{1}{2\pi} \int_{-\infty}^{+\infty} e^{i\lambda x} \, d\lambda \quad (3\text{-}25)$$

For liquids and gases under ordinary conditions, the correlation length of the density–density correlation function evaluated at equal times is much smaller than the optical wavelength,[†] which amounts to saying that $e^{i\mathbf{k}_1\cdot\mathbf{r}} \simeq 1$ in Eq. (3-24), since $k_1 \leqslant 2k_0$. Thus it follows from Eq. (3-24) that

$$S(-\mathbf{k}_1) \simeq S(0) = \int \langle \rho_1(\mathbf{r}', 0) \rho_1(0,0) \rangle \, d\mathbf{r}' \quad (3\text{-}26)$$

that is, the angular distribution of the total intensity does not depend, apart from the geometrical factor $\sin^2 \gamma$, on the scattering angle.

[†] We observe that this is not always the case for the correlation function at different times due to the transport of fluctuations, which can occur at sound velocity, thus providing a spatial correlation on a larger scale (Landau and Lifshitz, 1960, Chapter XIV).

The quantity $S(0)$ can now be evaluated according to the following thermodynamical argument. Let us consider a volume element V_1 large compared to the cube of the equal time correlation length, but small compared to the scattering volume V_{sc}. If N is the total number of molecules in V_1, we have

$$\langle N^2 \rangle = \left\langle \left[\int_{V_1} \rho(\mathbf{r}')\, d\mathbf{r}' \right]^2 \right\rangle$$

$$= \int_{V_1} \int_{V_1} \langle \rho(\mathbf{r}')\rho(\mathbf{r}'') \rangle\, d\mathbf{r}'\, d\mathbf{r}''$$

$$= \rho_0^2 V_1^2 + V_1 \int \langle \rho_1(\mathbf{r}')\rho_1(0) \rangle\, d\mathbf{r}' \qquad (3\text{-}27)$$

where the temporal argument has been dropped due to the stationarity hypothesis. According to Eq. (3-27) the mean-square fluctuation of N is given by

$$\frac{\langle N^2 \rangle - \langle N \rangle^2}{V_1} = \int \langle \rho_1(\mathbf{r}')\rho_1(0) \rangle\, d\mathbf{r}' = S(0) \qquad (3\text{-}28)'$$

The smallness hypothesis on V_1 allows us to consider it as a system in contact with a large heat reservoir given by the remaining fluid. Therefore, it is possible to apply a well-known thermodynamical relation that gives (Landau and Lifshitz, 1958, Chapter XII)

$$S(0) = \frac{\langle N^2 \rangle - \langle N \rangle^2}{V_1} = \rho_0^2 K_B T \chi_T \qquad (3\text{-}29)$$

If one now substitutes Eq. (3-29) into Eq. (3-23), having taken Eq. (3-26) into account, one obtains the following expression for the total scattering intensity per unit solid angle:

$$I(\eta) = \frac{V_{sc} I_0}{(4\pi r)^2} k_0^4 \sin^2 \gamma \left(\frac{d\varepsilon}{d\rho} \right)^2 \rho_0^2 K_B T \chi_T \qquad (3\text{-}30)$$

The extinction coefficient h can then be evaluated by its definition

$$h = \frac{1}{V_{sc} I_0} \int I(\mathbf{\eta}) \, r^2 \, d\Omega \qquad (3\text{-}31)$$

where Ω is the solid angle. In fact, by introducing a system of spherical coordinates with polar axis in the direction of $\mathbf{E}_0$, one has

$$\int_0^{2\pi} d\varphi \int_0^\pi \sin^2 \gamma \sin \gamma \, d\gamma = \frac{8\pi}{3} \qquad (3\text{-}32)$$

so that

$$h = \frac{\omega_0^4}{6\pi c^4} K_B T \rho_0^2 \chi_T \left(\frac{d\varepsilon}{d\rho} \right)^2_{\rho = \rho_0} \qquad (3\text{-}33)$$

which is Einstein's formula in the case of ε independent of T. Of course, one can treat the temperature-dependent case as well taking into account the explicit temperature fluctuations, thus obtaining the complete Einstein formula given by Eq. (3-6).

In conclusion, the previous treatment furnishes a macroscopic expression for the spectrum of light scattered in a given direction due to dielectric constant fluctuations, which in particular allows one to recover Einstein's formula for the total scattered intensity. In order to show the limitations of this phenomenological treatment, one has to compare the results of this section with the corresponding ones obtained by means of the microscopic approach.

III.3 Comparison with the First-Order Microscopic Treatment

The microscopic expression of the total scattered intensity per unit solid angle is obtained, in the single-scattering approximation, by integrating Eq. (2-53) with respect to frequency.

This immediately yields

$$I(\eta) = \int_{-\infty}^{+\infty} I(\omega, \eta) \, d\omega$$

$$= 2\pi Z^2 \int d\mathbf{r}' \, d\mathbf{r}'' \exp[i\mathbf{k}_1 \cdot (\mathbf{r}' - \mathbf{r}'')] \langle n_1(\mathbf{r}', 0) n_1(\mathbf{r}'', 0) \rangle \quad (3\text{-}34)$$

where $n_1(\mathbf{r}, t)$ is the microscopic density as defined in Eq. (2-27). By repeating the same procedure of the preceding section one obtains from Eq. (3-34)

$$h = \frac{8\pi}{3} \frac{\omega_0^4}{c^4} \alpha^2 \frac{\langle N^2 \rangle - \langle N \rangle^2}{V_1} \quad (3\text{-}35)$$

which has to be compared with Eq. (3-33). The two expressions coincide if

$$\left(\frac{d\varepsilon}{d\rho} \right)_{\rho = \rho_0} = 4\pi\alpha \quad (3\text{-}36)$$

which is a relation consistent with the Lorenz–Lorentz equation given by Eq. (3-7) to the lowest significant order in $\alpha\rho_0$. This means that the Einstein–Smolouchosky formula possesses a microscopic justification, at least to this order and in the single-scattering approximation, provided that the fluctuating dielectric constant $\varepsilon(\mathbf{r}, t)$ is related to the local density by the same equation connecting the dielectric constant ε_0 (as expressed in terms of ensemble average) to the average density.

The simple situation of an ideal gas can, in particular, be treated by using either Eq. (3-33) or (3-35). In the first case one has to take into account that

$$\left(\frac{\partial \rho}{\partial p} \right)_{T, \rho = \rho_0} = \frac{1}{K_B T}$$

while in the second it is possible to take advantage of the relation $\langle N^2 \rangle - \langle N \rangle^2 = \langle N \rangle$ following from the Poissonian distribution of the number of independent particles in a small volume (Chandrasekhar, 1943). Both calculations yield the Rayleigh equation expressed by Eq. (3-5).

The proof of the equivalence of the microscopic and macroscopic approaches is completed by showing that

$$
\int_{V_{sc}} \int_{V_{sc}} d\mathbf{r}' \, d\mathbf{r}'' \exp[i\mathbf{k}_1 \cdot (\mathbf{r}' - \mathbf{r}'')] \langle \Delta n_1(\mathbf{r}', \tau) \Delta n_1(\mathbf{r}'', 0) \rangle
$$

$$
= \int_{V_{sc}} \int_{V_{sc}} d\mathbf{r}' \, d\mathbf{r}'' \exp[i\mathbf{k}_1 \cdot (\mathbf{r}' - \mathbf{r}'')] \langle \rho_1(\mathbf{r}', \tau) \rho_1(\mathbf{r}'', 0) \rangle
$$

$$(3\text{-}37)$$

This is immediately achieved by evaluating the integrals on both sides of Eq. (3-37) as a superposition of integrals over volumes much smaller than λ_0^3 but large compared with v (see Section III.2). In fact, the left-hand side of Eq. (3-37) can be rewritten as

$$
\left\langle \sum_i \sum_j \exp[i\mathbf{k}_1 \cdot (\mathbf{r}_i - \mathbf{r}_j)] \int_{V_i} \int_{V_j} d\mathbf{r}' \, d\mathbf{r}'' \Delta n_1(\mathbf{r}', \tau) \Delta n_1(\mathbf{r}'', 0) \right\rangle
$$

$$
= \left\langle \sum_i \sum_j \exp[i\mathbf{k}_1 \cdot (\mathbf{r}_i - \mathbf{r}_j)] \Delta N_1(V_i, \tau) \Delta N_1(V_j, 0) \right\rangle \qquad (3\text{-}38)
$$

where $\mathbf{r}_i$ is the position of a point inside V_i and $\Delta N_1(V_i, \tau)$ the fluctuation of the particle number contained in V_i at time τ. The same procedure applies on the right-hand side of Eq. (3-37) by obtaining the same result, after observing that

$$
\int_{V_i} \rho_1(\mathbf{r}', \tau) \, d\mathbf{r}' = \Delta N_1(V_i, \tau) \qquad (3\text{-}39)
$$

having neglected terms of the order v/V_i.

Thus one can conclude that, in the limit $\rho_0 \alpha \ll 1$, both treatments can be employed to give the same results for the extinction coefficient as well as for the spectrum as long as microscopic single-scattering contributions are taken into account.

The problem becomes much more complicated whenever $\rho_0 \alpha$ is no longer negligible with respect to unity and microscopic multiple scattering comes into play. While no satisfactory expression of the spectrum of the multiply scattered radiation exists, the Einstein–Smolouchosky formula can also be shown to maintain its validity if a certain number of higher-order contributions is taken into account.

III.4 Microscopic Extension of the Einstein–Smolouchosky Formula

A molecular derivation of the Einstein–Smolouchosky formula can be performed in a way that includes multiple–scattering contributions. The direct way of doing it is to evaluate the total scattered intensity by starting from the microscopic expression of the scattered field obtained by extending the calculation performed in Section I.6 for the single-scattering case; the extinction coefficient h then has to be related to the dielectric constant ε_0, after this has been calculated by including the same multiple-scattering terms considered in evaluating the scattered field.

This approach is the one followed by Fixman (1955), who has taken into account microscopic triple-scattering contributions by considering the series in Eq. (1-84) to include the terms in α^3, thus extending the previous derivations of Yvon (1937) and Zimm (1945).

A different method, which can be developed in particular in the framework of scattering theory (Rosenfeld, 1951; Bullough, 1968), is based on the derivation, starting from Eq. (1-82), of an integral equation for the polarization as defined by Eq. (3-2).

One starts by using the set of Eqs. (1-82) in the form

$$\hat{\mathbf{P}}_i = \alpha(\omega_0)\left[\hat{\mathbf{E}}_e(\mathbf{R}_i, t) + \sum_{j=1, j \neq i}^{N} \mathbf{T}(\mathbf{R}_i, \mathbf{R}_j) \cdot \hat{\mathbf{P}}_j \right] \quad (3\text{-}40)$$

where $\hat{\mathbf{E}}_e(\mathbf{r}, t)$ is the electric field incident from outside and $\mathbf{R}_i$ the position of the ith molecule at time t. Equations (3-40) can be formally solved by iteration, thus obtaining

$$\hat{\mathbf{P}}_i = \alpha \hat{\mathbf{E}}_e(\mathbf{R}_i, t) + \alpha^2 \sum_{j=1, j \neq i}^{N} \mathbf{T}(\mathbf{R}_i, \mathbf{R}_j) \cdot \hat{\mathbf{E}}_e(\mathbf{R}_j, t)$$

$$+ \alpha^3 \sum_{j=1, j \neq i}^{N} \sum_{k=1, k \neq j}^{N} \mathbf{T}(\mathbf{R}_i, \mathbf{R}_j) \cdot \mathbf{T}(\mathbf{R}_j, \mathbf{R}_k) \cdot \hat{\mathbf{E}}_e(\mathbf{R}_k, t) + \cdots \tag{3-41}$$

and, after multiplying each equation by the factor $\delta(\mathbf{r} - \mathbf{R}_i)$ and summing,

$$\sum_{i=1}^{N} \delta(\mathbf{r} - \mathbf{R}_i) \hat{\mathbf{P}}_i$$

$$= \alpha \sum_{i=1}^{N} \delta(\mathbf{r} - \mathbf{R}_i) \hat{\mathbf{E}}_e(\mathbf{r}, t) + \alpha^2 \int_{V_{sc}} d\mathbf{r}' \, \mathbf{T}(\mathbf{r}, \mathbf{r}')$$

$$\times \sum_{i=1}^{N} \sum_{\substack{j=1 \\ j \neq i}}^{N} \delta(\mathbf{r} - \mathbf{R}_i) \delta(\mathbf{r}' - \mathbf{R}_j) \cdot \hat{\mathbf{E}}_e(\mathbf{r}', t) + \alpha^3 \int_{V_{sc}} d\mathbf{r}'$$

$$\times \int_{V_{sc}} d\mathbf{r}'' \, \mathbf{T}(\mathbf{r}, \mathbf{r}') \cdot \mathbf{T}(\mathbf{r}', \mathbf{r}'') \sum_{j=1, j \neq i}^{N} \sum_{k=1, k \neq j}^{N} \delta(\mathbf{r} - \mathbf{R}_i)$$

$$\times \delta(\mathbf{r}' - \mathbf{R}_j) \delta(\mathbf{r}'' - \mathbf{R}_k) \cdot \hat{\mathbf{E}}_e(\mathbf{r}'', t) + \cdots \tag{3-42}$$

Taking the ensemble average of both sides of Eq. (3-42) and recalling the relations given in Eqs. (2-25)–(2-28) (and the corresponding ones that hold for successive orders), one immediately sees that the polarization $\hat{\mathbf{P}}(\mathbf{r}, t)$ can be expressed as

$$\hat{\mathbf{P}}(\mathbf{r}, t) = \alpha N f_1(\mathbf{r}) \hat{\mathbf{E}}_e(\mathbf{r}, t) + \alpha^2 N(N-1) \int_{V_{sc}} d\mathbf{r}' \, \mathbf{T}(\mathbf{r}, \mathbf{r}')$$

$$\times f_2(\mathbf{r}, \mathbf{r}') \cdot \hat{\mathbf{E}}_e(\mathbf{r}, t) + \alpha^3 N(N-1)(N-2) \int_{V_{sc}} d\mathbf{r}' \int_{V_{sc}} d\mathbf{r}''$$

$$\times \mathbf{T}(\mathbf{r}, \mathbf{r}') \cdot \mathbf{T}(\mathbf{r}', \mathbf{r}'') f_3(\mathbf{r}, \mathbf{r}', \mathbf{r}'') \cdot \hat{\mathbf{E}}_e(\mathbf{r}'', t) + \cdots \tag{3-43}$$

In particular, if one assumes the absence of any correlation between the N particles $(N \gg 1)$ of the system, Eq. (3-43) is equivalent, in the homogeneous case, to the integral equation

$$\hat{\mathbf{P}}(\mathbf{r}, t) = \alpha \left\{ \rho_0 \, \hat{\mathbf{E}}_e(\mathbf{r}, t) + \rho_0 \int_{V_{sc}} d\mathbf{r}' \, \mathbf{T}(\mathbf{r}, \mathbf{r}') \cdot \hat{\mathbf{P}}(\mathbf{r}', t) \right\} \quad (3\text{-}44)$$

where the relation $f_n = V_{sc}^{-n}$ has been used.

By taking into account the presence of a two-particle correlation, Eq. (3-43) yields (see Bullough, 1968)[†]

$$\hat{\mathbf{P}}(\mathbf{r}, t) = \alpha \left\{ \rho_0 \, \hat{\mathbf{E}}_e(\mathbf{r}, t) + \rho_0 \, V_{sc}^2 \int d\mathbf{r}' \, \mathbf{T}(\mathbf{r}, \mathbf{r}') \, f_2(\mathbf{r}, \mathbf{r}') \cdot \hat{\mathbf{P}}(\mathbf{r}', t) \right\}$$

$$(3\text{-}45)$$

The introduction of an integral equation for the dipole moment density has been recognized for a long time (Darwin, 1924) as the starting point for understanding from a microscopic point of view the problem of electromagnetic propagation in matter. In particular, this formulation allows us to give a microscopic interpretation of the mechanism by means of which an electromagnetic wave incident from vacuum on a material system is extinguished by interference with a part of the field generated by the induced dipoles and is replaced by a new wave propagating with a different velocity smaller than c. This statement is known as the *Ewald–Oseen extinction theorem* (Born and Wolf, 1970; Chapter II); for a macroscopic interpretation of this theorem see the work of Pattanayak and Wolf (1972).

The method based on the use of an integral equation for the dipole moment density has been developed in order to obtain an expression for the refractive index in a number of physical situations (Hoek, 1941; Mazur and Mandel, 1956; Mazur and Terwiel, 1964).

[†] We note that the correspondence between Eqs. (3-45) and (1-2) of Bullough's paper is achieved by observing that the latter equation contains the average dipole moment per particle and a two-particle distribution function $g(\mathbf{r}, \mathbf{r}')$, such that $g(\mathbf{r}, \mathbf{r}') = V_{sc}^2 f_2(\mathbf{r}, \mathbf{r}')$.

In the framework of scattering, the imaginary part of the complex refractive index is directly related to the extinction coefficient. The analytical procedure followed in order to solve the integral equation is to look for a trial solution $\hat{\mathbf{P}}(\mathbf{r}, t)$ satisfying the wave equation (Rosenfeld, 1951; Chapter VI)

$$\nabla^2 \hat{\mathbf{P}}(\mathbf{r}, t) + m^2 k_0^2 \, \hat{\mathbf{P}}(\mathbf{r}, t) = 0 \tag{3-46}$$

together with the transversality condition

$$\nabla \cdot \hat{\mathbf{P}}(\mathbf{r}, t) = 0 \tag{3-47}$$

derived from the assumption that no sources are present in the medium. The simplest solution of both Eqs. (3-46) and (3-47) and of the integral equation is (Bullough, 1967)

$$\hat{\mathbf{P}}(\mathbf{r}, t) = \mathbf{u} P_0 \exp(i m k_0 \cdot \mathbf{r} - i\omega_0 t) \tag{3-48}$$

if the incident field is given by

$$\hat{\mathbf{E}}_e(\mathbf{r}, t) = \mathbf{u} A_0 \exp(i k_0 \cdot \mathbf{r} - i\omega_0 t) \tag{3-49}$$

so that one has to look for an appropriate value of the parameter m. The behavior of the electric field $\hat{\mathbf{E}}(\mathbf{r}, t)$ inside the medium is accordingly of the form

$$\hat{\mathbf{E}}(\mathbf{r}, t) = \mathbf{u} A_0 \exp(i m k_0 \cdot \mathbf{r} - i\omega_0 t) \tag{3-50}$$

which permits us to regard m as the (complex) refractive index of the medium. Equation (3-50) shows that the electromagnetic flux density is reduced by a factor $\exp(-2k_0 x\delta)$ after traveling a distance x, if

$$m = \mu + i\delta \tag{3-51}$$

Thus, since the extinction coefficient h can be considered equivalent (see Section III.5) to the relative decreasing of flux density per unit length of the scattering medium, one has

$$h = 2k_0 \delta \tag{3-52}$$

If one applies the above procedure to Eq. (3-44), one finds that m obeys the Lorenz–Lorentz equation

$$\frac{m^2 - 1}{m^2 + 2} = \frac{4\pi}{3} \alpha \rho_0 \tag{3-53}$$

so that it is real and does not account for any scattered field. This is not surprising since Eq. (3-44) describes in a correct way only a homogeneous system in which no density fluctuations take place (as a crystal), so that the scattered microfields interfere in a destructive way. The situation is different if one considers the cases described by Eq. (3-45) or, more generally, by Eq. (3-43), in which the correlation between the scattering molecules is taken into account. Equation (3-45) leads to the following expression for the refractive index (Bullough, 1967):

$$\frac{m^2 - 1}{m^2 + 2} = \frac{4\pi}{3} \rho_0 \alpha [1 - \alpha \rho_0 J_1(m\mathbf{k}_0)]^{-1} \tag{3-54}$$

where

$$J_1(m\mathbf{k}_0) = \mathbf{u} \cdot \int_{V_{sc} - v} d\mathbf{r}' \, \mathbf{T}(\mathbf{r}, \mathbf{r}') \cdot \mathbf{u} \exp[im\mathbf{k}_0 \cdot (\mathbf{r} - \mathbf{r}')]$$
$$\times [V_{sc}^2 f_2(\mathbf{r}, \mathbf{r}') - 1] \tag{3-55}$$

Here v is a small excluded volume and the limit $v \to 0$ has to be taken.

The general Eq. (3-43) yields an expression for m (Bullough, 1967)

$$\frac{m^2 - 1}{m^2 + 2} = \frac{4\pi}{3} \rho_0 \alpha \left[1 - \sum_{r=1}^{\infty} (\rho_0 \alpha)^r J_r(m\mathbf{k}_0) \right]^{-1} \tag{3-56}$$

where the J_r's are complicated functionals of the correlation functions up to the rth order. Both Eqs. (3-54) and (3-56) give a complex value of m, thus accounting for the scattering process. In particular, Eq. (3-54) takes into account only single-scattering contributions, while Eq. (3-56) includes terms of all orders, as can be seen explicitly by comparing, order by order in $\alpha \rho_0$, the expressions of the extinction coefficient obtained by evaluating the envelope of scattered intensity starting from Eq. (1-84) and that derived by the present method (Rosenfeld, 1951, Chapter VI; Bullough and Hynne, 1968).

This approach also allows us to check in a direct way the validity of the Einstein–Smolouchosky equation after evaluating $\mu = \sqrt{\varepsilon}$ and δ. Referring to Eq. (3-54), the expression one arrives at for h is (Bullough, 1967)

$$h = \frac{\omega_0^4}{6\pi c^4} K_B T\rho_0^2 \chi_T \left(\frac{d\mu^2}{d\rho}\right)^2_{\rho=\rho_0} \frac{9}{\mu(\mu^2+2)^2} \qquad (3\text{-}57)$$

which differs from Eq. (3-33) by the factor $9/\mu(\mu^2+2)^2$. This discrepancy does not necessarily mean that the phenomenological result is incorrect. In fact, Eq. (3-57) has been derived by considering single scattering only, that is, by neglecting the contributions due to J_r for $r > 1$. Since this is valid when $\alpha\rho_0 \ll 1$, which implies $\mu \simeq 1$, one has $9/\mu(\mu^2+2)^2 \simeq 1$. If one considers now Eq. (3-56), corresponding to the case in which higher-order terms J_r cannot be neglected, one finds that the factor $9/\mu(\mu^2+2)^2$ is really a fictitious one and disappears when terms up to J_3, including four-particle correlations, are taken into account. More precisely, neglecting terms of order $(\alpha\rho_0)^4$ and $k_0^2 l_c^2$ (where l_c is an order of magnitude for the correlation length of the scattering medium), the extinction coefficient is given by (Bullough and Hynne, 1968; Bullough et al., 1968)

$$h = \frac{\omega_0^4}{6\pi c^4} K_B T\rho_0^2 \chi_T \left(\frac{d\mu^2}{d\rho}\right)^2_{\rho=\rho_0} \qquad (3\text{-}58)$$

which is the Einstein–Smolouchosky result.

III.5 Usefulness and Limitations of the Macroscopic Treatment

The identity between the expression of the extinction coefficient worked out by means of the fairly complete multiple-scattering microscopic evaluation and that obtained in the frame of the simple macroscopic approach suggests also that the macroscopic expression of the spectrum derived in Section (III.2) possesses a range of validity larger than that pertaining to the first-order microscopic treatment. More precisely, one should verify whether the expression of the scattered spectrum given by Eq. (3-21) possesses a more general validity than the molecular result given by Eq. (2-56), once multiple scattering contributions are taken into account in evaluating $d\varepsilon/d\rho$.

This comparison is actually performed in Section III.3, in the single-scattering approximation, provided that $\rho_0 \alpha \ll 1$, a case in which the two approaches are completely equivalent. Unfortunately, the evaluation of the spectrum is of increasing complexity when multiple-scattering terms have to be included, and a direct comparison has not yet been achieved. Of course, the validity of the macroscopic treatment for the extinction coefficient, once multiple molecular scattering up to neglected terms of the order $(\rho_0 \alpha)^4$ and $k_0^2 l^2$ is included, represents only a necessary condition for its applicability in calculating the spectrum. Another necessary condition is given by the absence of depolarized scattered radiation, since this is the result of the macroscopic approach, a fact that has to be experimentally tested case by case. In this respect, it is worth mentioning the result obtained by Gelbart (1972), who evaluates the depolarization factor Δ for light scattered by gases at high pressure taking into account three- and four-particle correlations [that is, up to neglected terms in $(\rho_0 \alpha)^4$]. The factor Δ turns out to be reduced with respect to that calculated by including only two-particle correlation [that is, up to neglected terms in $(\rho_0 \alpha)^2$] (see Fig. 3.1).[†]

[†] We observe that this cancellation mechanism is particularly efficient at liquid densities, as already noted in Section I.5 when considering the field exerted on a given molecule by the surrounding ones. The contribution arising from distant molecules becomes important near enough the critical point, where Δ has been shown to increase roughly as the square of the correlation length (Oxtoby and Gelbart, 1974a).

This result, together with the validity of the Einstein–Smolouchosky equation, seems to indicate that including second- and third-order scattering terms in a microscopic calculation tends to restore the validity of the macroscopic result, which is lost in the single-scattering microscopic approach (unless $\rho_0 \alpha \ll 1$). Although the approximation under consideration applies to most practical situations, it is useful to observe that the macroscopic treatment fails if terms of the order of $(\rho_0 \alpha)^5$ are taken into account. In this case one has (Bullough

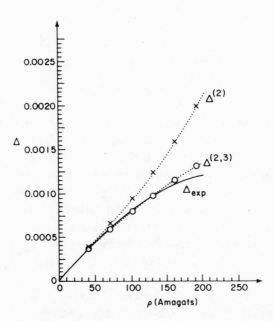

Fig. 3.1   *Depolarization factor* Δ *of light by argon gas at room temperature. Here* $\Delta^{(2)}$ *is the theoretical value taking into account only isolated pair interactions, while* $\Delta^{(2,3)}$ *includes the effects of three-particle correlations as well (after Gelbart, 1972).*

and Hynne, 1968)

$$h = \frac{\omega_0^4}{6\pi c^4} K_B T \rho_0^2 \chi_T \left(\frac{d\mu^2}{d\rho}\right)^2_{\rho=\rho_0} \frac{3+6\Delta}{3-4\Delta} \qquad (3\text{-}59)$$

where the *Cabannes factor* $(3+6\Delta)/(3-4\Delta)$ (Fabelinskii, 1968, Chapter I) modifying Eq. (3-58) is due to the depolarization of the scattered radiation connected with multiple scattering. At higher order in $\rho_0 \alpha$ the expression for h furnished by Eq. (3-59) fails.

It is worth noting that Eq. (3-50) is consistent with an exponential behavior of the intensity propagating in the medium

$$I(x) = I(0) e^{-hx} \qquad (3\text{-}60)$$

where x is the distance traveled, while the definition of h given in Eq. (3-31) is consistent with

$$I(x) = I(0)(1 - hx) \qquad (3\text{-}61)$$

The two equations are equivalent only for small values of hx or, equivalently, for distances such that the relative decrease in intensity is small with respect to unity. In the fundamental approach just described, the validity of the above approximation is tacitly assumed by neglecting the multiple-scattering contributions to h, which are not independent of the scattering volume, depending on both its magnitude and shape. Therefore, the treatment correctly describes situations in which the intensity of the transmitted beam is not strongly affected by scattering. From a microscopic point of view, this circumstance is usually referred to as a *single-scattering* condition (although, as we have seen, microscopic multiple-scattering contributions play a relevant role). This also corresponds to the fact that the macroscopic Maxwell's equations of Section (III.2) have been solved to a first-order approximation [see Eqs. (3-11)]. In this respect, it is worth noting that, since the only justification of these phenomenological equations lies in the possible comparison with the microscopic calculations, there is no *a priori* justification for iterating the procedure to higher orders in ε_1.

III.6   The Spectrum of Light Scattered by Normal Liquids

The spectral distribution $I(\omega, \eta)$ of light scattered by a fluid composed of optically isotropic molecules as furnished by Eq. (2-56) or (3-21) is expressed in terms of the correlation functions of the density fluctuations of the scattering medium. Thus to obtain its explicit form requires an investigation of the behavior of these correlation functions. Depending on whether the microscopic or macroscopic description is adopted, one deals, respectively, with the microscopic and macroscopic correlation functions $\langle \Delta n_1(\mathbf{r}, t) \Delta n_1(0,0) \rangle$ and $\langle \rho_1(\mathbf{r}, t) \rho_1(0,0) \rangle$. Use of the first one requires consideration of the BBGKY hierarchy, which contains the complete microscopic description of the system. The adoption of the second approach may offer great advantages for analytical simplicity whenever one deals with the *hydrodynamic regime*. It is characterized by the limit $qr_0 \ll 1$, where r_0 is the molecular collision mean free path and q a typical wavevector pertaining to the Fourier space transform of the quantities under study. This means that the hydrodynamical treatment allows us to study the behavior of the significant physical quantities over spatial scales larger than r_0. In our case, the largest wavevector involved is $2\mathbf{k}_0$, corresponding to backscattering, so that the condition is largely satisfied for liquids. This is not true for rarefied gases, which require a microscopic treatment, an example of which will be shown in Chapter V when studying light scattering from an electron gas.

Let us consider here the hydrodynamical approach applied to a simple fluid characterized by small fluctuations. The problem is to write a convenient set of equations for the significant fluctuating quantities by means of which one is able to derive the correlation $\langle \rho_1(\mathbf{r}, t)\rho_1(0,0)\rangle$.

To this end, one starts by writing the usual equations of fluid dynamics, expressing the conservation of mass, momentum, and energy in the form (Hunt, 1957)[†]

$$\frac{\partial \rho}{\partial t} + \nabla \cdot (\rho\mathbf{v}) = 0 \qquad (3\text{-}62)$$

$$\rho\frac{\partial \mathbf{v}}{\partial t} + \rho(\mathbf{v}\cdot\nabla)\mathbf{v} = -\nabla p + \nabla\sigma \qquad (3\text{-}63)$$

$$\rho\frac{\partial \varepsilon}{\partial t} + \rho(\mathbf{v}\cdot\nabla)\varepsilon = \sigma:\mathbf{d} - \nabla\cdot\mathbf{q} \qquad (3\text{-}64)$$

where ρ, $\mathbf{v}$, p, and ε are, respectively, the fluid *density*, *velocity*, *pressure*, and *specific energy per unit volume*, and $\mathbf{d}$ is the *rate of deformation tensor* defined by

$$d_{ij} = \frac{1}{2}\left(\frac{\partial v_i}{\partial x_j} + \frac{\partial v_j}{\partial x_i}\right) \qquad (3\text{-}65)$$

The quantities $\mathbf{q}$ and σ are, respectively, the *heat flux* and the *stress tensor* given by

$$\mathbf{q} = -\lambda\nabla T \qquad (3\text{-}66)$$

where λ is the *thermal conductivity* and T the absolute temperature, and

$$\sigma_{ij} = (\eta_b - \tfrac{2}{3}\eta_{sh})\sum_k d_{kk}\delta_{ij} + 2\eta_{sh}d_{ij} \qquad (3\text{-}67)$$

with η_{sh} and η_b the *shear* and *bulk viscosity* and δ_{ij} the usual Kronecker symbol.

These are the classical deterministic hydrodynamic equations,

---

[†] The symbol $\nabla\sigma$ in Eq. (3-63) indicates the vector whose *i*th component is $\sum_k \partial\sigma_{ik}/\partial x_k$. For the definition of the scalar product $\sigma:\mathbf{d}$ see the footnote on p. 102.

which do not take into account the stochastic nature of the fluid arising from the discrete structure of matter. In statistical language, this amounts to saying that Eqs. (3-62)–(3-64) describe the behavior of ensemble-averaged quantities. The set of equations describing the result of each single experiment, that is, the *fluctuating instantaneous quantities*, is obtained, according to Landau and Lifshitz (1959, Chapter XVII) by adding a purely random stress tensor and a purely random heat flux vector to σ_{ik} and $\mathbf{q}$. These equations can be written, for small fluctuations, in the linearized form

$$\frac{\partial A_\alpha(\mathbf{r}, t)}{\partial t} = -\sum_\beta \Gamma_{\alpha\beta} A_\beta(\mathbf{r}, t) + f_\alpha(\mathbf{r}, t) \qquad (3\text{-}68)$$

where $\Gamma_{\alpha\beta}$ are linear differential operators and

$$\{A_\alpha\} = \{\rho(\mathbf{r}, t) - \rho_0,\ \mathbf{v}(\mathbf{r}, t) - \mathbf{v}_0,\ T(\mathbf{r}, t) - T_0\}$$

Here ρ_0, $\mathbf{v}_0$, and T_0 indicate the ensemble averages, which are, in the particular case of homogeneous and stationary fluids, independent of space and time. We observe that p and ε no longer appear in Eqs. (3-68), since they can be expressed in terms of the other quantities by means of suitable thermodynamic relations (Hunt, 1957). The f_α's play the role of *random forces* corresponding to the random stress tensor and flux vector.

It is obvious that Eqs. (3-68) are, except for the presence of the f_α's, formally equivalent to the nonfluctuating hydrodynamic equations, derived by linearizing Eqs. (3-62)–(3-64) in the deterministic quantities $\rho_1(\mathbf{r}, t)$, $\mathbf{v}_1(\mathbf{r}, t)$, $T_1(\mathbf{r}, t)$, which represent the small differences of ρ, $\mathbf{v}$, and T from a value constant in space and time. The set of Eqs. (3-68) can be used to calculate the ensemble averages $\langle A_\alpha(\mathbf{r}, t) A_\gamma(0, 0) \rangle$, since from them one deduces

$$\frac{\partial}{\partial t} \langle A_\alpha(\mathbf{r}, t) A_\gamma(0, 0) \rangle = -\sum_\beta \Gamma_{\alpha\beta} \langle A_\beta(\mathbf{r}, t) A_\gamma(0, 0) \rangle$$
$$+ \langle f_\alpha(\mathbf{r}, t) A_\gamma(0, 0) \rangle \qquad (3\text{-}69)$$

These equations simplify considerably by observing that

$$\langle f_\alpha(\mathbf{r}, t)\, A_\gamma(0,0)\rangle = 0 \qquad (3\text{-}70)$$

which follows from the assumption that the f_α's vary in time much faster than the variables A_γ, so that one can reasonably assume

$$\langle f_\alpha(\mathbf{r}, t)\, A_\gamma(0,0)\rangle = \langle f_\alpha(\mathbf{r}, t)\rangle \langle A_\gamma(0,0)\rangle = 0 \qquad (3\text{-}71)$$

The solution of the resulting equations

$$\frac{\partial}{\partial t}\langle A_\alpha(\mathbf{r}, t)\, A_\gamma(0,0)\rangle = -\sum_\beta \Gamma_{\alpha\beta}\langle A_\beta(\mathbf{r}, t)\, A_\gamma(0,0)\rangle \qquad (3\text{-}72)$$

is equivalent to the one obtained by supposing that the fluctuations A_α obey the nonfluctuating hydrodynamic equations

$$\frac{\partial A_\alpha(\mathbf{r}, t)}{\partial t} = -\sum_\beta \Gamma_{\alpha\beta}\, A_\beta(\mathbf{r}, t) \qquad (3\text{-}73)$$

and by taking the ensemble average of its product with $A_\gamma(0,0)$. It is worth stressing that Eqs. (3-73) can be applied for evaluating the correlation functions $\langle A_\alpha(\mathbf{r}, t)\, A_\gamma(0,0)\rangle$ by virtue of Eq. (3-70), although the correct behavior of the A_α's is described by Eqs. (3-68). The outlined procedure above is equivalent to the so-called *Onsager hypothesis* of the regression of fluctuations, which states that a small statistical fluctuation existing at time $t = 0$ evolves according to the linearized hydrodynamics equation (Onsager, 1931).

The hydrodynamical approach to scattering problems, first proposed by Landau and Placzek (1934), has been treated in detail by Mountain (1966) (see also Rytov, 1957, and Foch, 1968; for recent theoretical approaches see Pike, 1974, and Lallemand, 1974).

The linearized hydrodynamic Eqs. (3-73) possess the explicit form

$$\frac{\partial \rho_1}{\partial t} + \rho_0 \nabla \cdot \mathbf{v} = 0 \quad (3\text{-}74)$$

$$\rho_0 \frac{\partial \mathbf{v}}{\partial t} + \frac{c_0^2}{\gamma} \nabla \rho_1 + \frac{c_0^2 \beta \rho_0}{\gamma} \nabla T_1 - \left(\frac{4}{3} \eta_{\text{sh}} + \eta_{\text{b}} \right) \nabla (\nabla \cdot \mathbf{v}) = 0 \quad (3\text{-}75)$$

$$\rho_0 c_{\text{v}} \frac{\partial T_1}{\partial t} + \frac{c_{\text{v}}(\gamma - 1)}{\beta} \frac{\partial \rho_1}{\partial t} - \lambda \nabla^2 T_1 = 0 \quad (3\text{-}76)$$

where c_0 is the low-frequency limit of the *sound velocity*, β the *thermal expansion coefficient*, γ the ratio between the *specific heats* at constant pressure and volume c_{p} and c_{v}, and it has been assumed that $\mathbf{v}_0 = 0$. The set of Eqs. (3-74)–(3-76) can be solved in a standard way by applying the Fourier (in space) and Laplace (in time) transform techniques. This last choice is the suitable one since we are interested in solving an initial-value problem. Let us define

$$\tilde{\rho}_1(\mathbf{k}, s) = \int d\mathbf{r} \int_0^\infty dt \exp(-i\mathbf{k} \cdot \mathbf{r} - st) \rho_1(\mathbf{r}, t) \quad (3\text{-}77)$$

$$\tilde{T}_1(\mathbf{k}, s) = \int d\mathbf{r} \int_0^\infty dt \exp(-i\mathbf{k} \cdot \mathbf{r} - st) T_1(\mathbf{r}, t) \quad (3\text{-}78)$$

$$\tilde{\rho}_1(\mathbf{k}) = \int d\mathbf{r} \, e^{-i\mathbf{k} \cdot \mathbf{r}} \rho_1(\mathbf{r}, 0) \quad (3\text{-}79)$$

$$\tilde{T}_1(\mathbf{k}) = \int d\mathbf{r} \, e^{-i\mathbf{k} \cdot \mathbf{r}} T(\mathbf{r}, 0) \quad (3\text{-}80)$$

One can eliminate the velocity by applying the divergence operator to Eq. (3-75) and taking advantage of Eq. (3-74). By taking the Fourier–Laplace transform of the resulting equation, as well as of Eq. (3-76), one obtains the system of two linear equations for the unknown quantities $\tilde{\rho}_1(\mathbf{k}, s)$ and $\tilde{T}_1(\mathbf{k}, s)$:

$$\tilde{\rho}_1(\mathbf{k}, s)\left[s^2 + \frac{c_0^2 k^2}{\gamma} + \left(\frac{4}{3}\eta_{\text{sh}} + \eta_{\text{b}}\right)\frac{k^2 s}{\rho_0}\right] + \frac{c_0^2 \beta \rho_0 k^2}{\gamma}\tilde{T}_1(\mathbf{k}, s)$$

$$= \tilde{\rho}_1(\mathbf{k})\left[s + \left(\frac{4}{3}\eta_{\text{sh}} + \eta_{\text{b}}\right)\frac{k^2}{\rho_0}\right] \qquad (3\text{-}81)$$

$$\tilde{\rho}_1(\mathbf{k}, s)\left[-sc_v\frac{(\gamma - 1)}{\beta}\right] + \tilde{T}_1(\mathbf{k}, s)(\rho_0 c_v s + \lambda k^2)$$

$$= -\tilde{\rho}(\mathbf{k})c_v\frac{(\gamma - 1)}{\beta} + \rho_0 c_v \tilde{T}_1(\mathbf{k}) \qquad (3\text{-}82)$$

This system can be solved for $\tilde{\rho}_1(\mathbf{k}, s)$, which turns out to be a linear combination of $\tilde{\rho}_1(\mathbf{k})$ and $\tilde{T}(\mathbf{k})$. Due to the statistical independence between density and temperature when taken at the same time (Landau and Lifshitz, 1958, Chapter XII), we can neglect in the solution the contribution of $\tilde{T}_1(\mathbf{k})$, since we are interested in evaluating the quantity $\langle \rho_1(\mathbf{r}, t)\rho_1(0, 0)\rangle$. Thus we can write

$$\tilde{\rho}_1(\mathbf{k}, s) = \frac{s^2 + (a+b)k^2 s + abk^4 + c_0^2[1 - (1/\gamma)]k^2}{s^3 + (a+b)k^2 s^2 + (c_0^2 k^2 + abk^4)s + a(c_0^2 k^4/\gamma)}\tilde{\rho}_1(\mathbf{k})$$

$$(3\text{-}83)$$

with

$$a = \lambda/(\rho_0 c_v) \quad \text{and} \quad b = \left(\tfrac{4}{3}\eta_{\text{sh}} + \eta_{\text{b}}\right)/\rho_0$$

In order to invert the Laplace transform of Eq. (3-77), one has to find the roots of the denominator of the right-hand side of Eq. (3-83), which is in the form of a cubic equation. Its solutions can be given in a simple approximated form, valid for a large class of scattering systems, to the lowest order in the parameters $ak^2/c_0 k$ and $bk^2/c_0 k$ as

$$s = \pm ic_0 k - \Gamma k^2, \quad s = -\frac{\lambda k^2}{\rho_0 c_{\text{p}}} \qquad (3\text{-}84)$$

having defined

$$\Gamma = \frac{1}{2}\left[\frac{\frac{4}{3}\eta_{sh} + \eta_b}{\rho_0} + \frac{1}{\rho_0}\left(\frac{\lambda}{c_v} - \frac{\lambda}{c_p}\right)\right] \tag{3-85}$$

The inversion of Eq. (3-77) yields, with the help of Eq. (3-84),

$$\rho_1(\mathbf{r}, t) = \frac{1}{(2\pi)^3}\int d\mathbf{k}\, e^{i\mathbf{k}\cdot\mathbf{r}}\left\{\frac{c_p - c_v}{c_p}\exp\left(-\frac{\lambda k^2}{\rho_0 c_p}t\right)\right.$$

$$\left. + \frac{c_v}{c_p}e^{-\Gamma k^2 t}\cos(c_0 kt)\right\}\tilde{\rho}_1(\mathbf{k}), \quad t > 0 \tag{3-86}$$

so that the correlation function of density fluctuations is given by

$$\langle\rho_1(\mathbf{r}, t)\rho_1(0, 0)\rangle = \frac{1}{(2\pi)^6}\int d\mathbf{k}\, d\mathbf{k}'\left\{\frac{c_p - c_v}{c_p}\exp\left(-\frac{\lambda k^2}{\rho_0 c_p}|t|\right)\right.$$

$$\left. + \frac{c_v}{c_p}e^{-\Gamma k^2|t|}\cos(c_0 kt)\right\}e^{i\mathbf{k}\cdot\mathbf{r}}\langle\tilde{\rho}_1(\mathbf{k})\tilde{\rho}_1(\mathbf{k}')\rangle,$$

$$\text{for every } t \tag{3-87}$$

having taken into account that the density correlation function is an even function of time, as a consequence of the stationarity hypothesis.

We now observe that the homogeneity and isotropy hypothesis, that is, the condition that the quantity $\langle\rho_1(\mathbf{r}', t)\rho_1(\mathbf{r}'', 0)\rangle$ depends on space only through the difference $|\mathbf{r}' - \mathbf{r}''|$, implies

$$\langle\tilde{\rho}_1(\mathbf{k})\tilde{\rho}_1(\mathbf{k}')\rangle = \int \exp(-i\mathbf{k}\cdot\mathbf{r} - i\mathbf{k}'\cdot\mathbf{r}')\langle\rho_1(\mathbf{r}, 0)\rho_1(\mathbf{r}', 0)\rangle\, d\mathbf{r}\, d\mathbf{r}'$$

$$= \int \exp[-i(\mathbf{k} + \mathbf{k}')\mathbf{r}' + i\mathbf{k}\cdot\boldsymbol{\xi}]\langle\rho_1(\boldsymbol{\xi}, 0)\rho_1(0, 0)\rangle$$

$$\times d\mathbf{r}'\, d\boldsymbol{\xi}$$

$$= (2\pi)^3\,\delta(\mathbf{k} + \mathbf{k}')\, S(k) \tag{3-88}$$

where

$$S(k) = \int e^{i\mathbf{k}\cdot\xi} \langle \rho_1(\xi,0)\rho_1(0,0)\rangle \, d\xi \tag{3-89}$$

By substituting Eq. (3-88) into Eq. (3-87) one immediately obtains

$$\langle \rho_1(\mathbf{r},t)\rho_1(0,0)\rangle = \frac{1}{(2\pi)^3}\int d\mathbf{k}\left\{\frac{c_p-c_v}{c_p}\exp\left(-\frac{\lambda k^2|t|}{\rho_0 c_p}\right)\right.$$
$$\left.+\frac{c_v}{c_p}\exp(-\Gamma k^2|t|)\cos(c_0 kt)\right\} e^{i\mathbf{k}\cdot\mathbf{r}} F(k)$$

$$\tag{3-90}$$

which, inserted in Eq. (3-22), gives

$$S(-\mathbf{k}_1,\omega-\omega_0)$$

$$= \frac{1}{\pi}\,\mathrm{Re}\int_0^\infty \exp[i(\omega-\omega_0)t]\left\{\frac{c_p-c_v}{c_p}\exp\left(-\frac{\lambda k_1^2}{\rho_0 c_p}t\right)\right.$$
$$\left.+\frac{c_v}{c_p}\exp(-\Gamma k_1^2 t)\cos(c_0 k_1 t)\right\} S(k_1)\,dt$$

$$= \frac{1}{2\pi}\left\{\frac{c_p-c_v}{c_p}\frac{2\lambda k_1^2/\rho_0 c_p}{(\omega-\omega_0)^2+(\lambda^2 k_1^4/\rho_0^2 c_p^2)}+\frac{c_v}{c_p}\right.$$

$$\times\left[\frac{\Gamma k_1^2}{(\Gamma k_1^2)^2+(\omega-\omega_0+c_0 k_1)^2}\right.$$

$$\left.\left.+\frac{\Gamma k_1^2}{(\omega-\omega_0-c_0 k_1)^2+(\Gamma k_1^2)^2}\right]\right\} S(k_1) \tag{3-91}$$

In this way, according to Eq. (3-21), the expression of the spectrum of the scattered intensity is evaluated except for the angle-dependent factor $S(k_1)$. This depends on the spontaneously produced fluctuation $\rho_1(0)$, so that its determination requires a suitable thermodynamical approach. We are dealing with fluids in ordinary conditions, so that all relevant information is contained in Eq. (3-29), since the smallness of the correlation length with respect to the wavelength implies $S(k_1) \simeq S(0)$. We shall examine in the next section the case in which this condition is not verified.

Inspection of Eq. (3-91) shows that the spectrum of the scattered light consists of one broadened Lorentzian unshifted line around the incident frequency ω_0 (*Rayleigh line*) and of two Lorentzian wings centered around the two frequencies $\omega_0 \pm c_0 k_1$ (*Mandel'shtam–Brillouin doublet*) (see Fig. 3.2).

One can gain direct physical insight into the process responsible for this behavior by recalling the way the dielectric constant was expressed in terms of two independent thermodynamic variables [see Eq. (3-20)], density and temperature. Another possible choice would have been to expand the fluctuating part of the dielectric constant $\varepsilon_1(\mathbf{r}, t)$ in terms of the fluctuating parts s_1 and p_1 of the statistically independent thermodynamic variables, entropy s and pressure p, as (Landau and Lifshitz, 1958, Chapter XII)

$$\varepsilon_1(\mathbf{r}, t) = \left(\frac{\partial \varepsilon}{\partial s}\right)_p s_1(\mathbf{r}, t) + \left(\frac{\partial \varepsilon}{\partial p}\right)_s p_1(\mathbf{r}, t) \qquad (3\text{-}92)$$

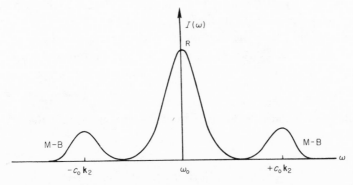

Fig. 3.2   *Spectrum of scattered light exhibiting the central unshifted Rayleigh line and the two symmetrical Mandel'shtam–Brillouin lines.*

According to Eqs. (3-19) and (3-92) the spectrum of the scattered light would have been expressed in terms of the correlation functions of the pressure and entropy fluctuations $\langle s_1(\mathbf{r}, t) s_1(0, 0) \rangle$ and $\langle p_1(\mathbf{r}, t) p_1(0, 0) \rangle$. In order to evaluate these quantities, one has to write the hydrodynamics equations for p_1 and s_1, which are analogous to the corresponding equations for ρ_1 and T_1. The structure of these equations is such that no coupling is present between s_1 and p_1 (Landau and Lifshitz, 1959; Chapter V), that is, the equation governing the evolution of p_1 does not contain s_1 and vice versa (this is only approximately true; see, for example, Cohen *et al.*, 1971). This in turn implies that the initial absence of correlation between p_1 and s_1 is maintained in time and that the cross-correlation terms $\langle s_1(\mathbf{r}, t) p_1(0, 0) \rangle$ and $\langle p_1(\mathbf{r}, t) s_1(0, 0) \rangle$ vanish in the hydrodynamical limit. Furthermore, pressure fluctuations at constant entropy propagate with sound velocity c_0 as damped waves so as to account for the Mandel'stham–Brillouin doublet. Entropy fluctuations at constant pressure, conversely, do not propagate and decay exponentially in time due to thermal conduction, thus being responsible for the central Rayleigh line.

We finally observe that modifications of the calculation leading to the expression of the scattered spectrum have been considered by employing refined forms of the hydrodynamic equations describing physical systems different from the singly relaxing fluid considered above (see, for example, Lebowitz *et al.*, 1969; for an excellent bibliography, see Fleury and Boon, 1973).

III.7 Some Remarks on Light Scattering from Liquids near the Critical Point

The method of light scattering can furnish particularly useful information whenever the fluid under investigation approaches its *critical point*. It is well known that, since the isothermal compressibility χ_T diverges when approaching the critical state, the density fluctuations become larger and larger so that the thermodynamical approach based on the smallness of fluctuations, leading to Eq. (3-29), is no longer valid. This is superseded by the classical treatment due to Ornstein and Zernike (1918) by means of which it is possible to obtain the equal-time density–density correlation function, which exhibits a long-range correlation in the neighborhood of the critical point. More precisely, one has (Landau and Lifshitz, 1958, Chapter XII)

$$\frac{\langle \rho_1(\mathbf{r},0)\,\rho_1(0,0)\rangle}{\rho_0^2} = A\,\frac{e^{-r/\xi}}{r} \tag{3-93}$$

with

$$A = \frac{\chi_T\,K_B\,T}{4\pi\xi^2} \tag{3-94}$$

where the correlation length ξ is proportional to $\chi_T^{1/2}$ and thus becomes very large in the critical region.

In order to determine the explicit expression for ξ, various microscopic treatments have been considered (Fixman, 1960; Lebowitz and Percus, 1963; van Kampen, 1964; Choy and Mayer, 1967). One has, for example (van Kampen, 1964),

$$\xi^2 = \tfrac{1}{2}\chi_T\,\rho_0^2\,w_2 \tag{3-95}$$

where

$$w_2 = \tfrac{1}{3}\int r^2\,w(r)\,d\mathbf{r} \tag{3-96}$$

with $w(r)$ the intermolecular attractive potential; A turns out to be of the order of the intermolecular force range.

The existence of a long-range correlation limits the use of the hydrodynamical equations for evaluating the temporal behavior of $\langle \rho_1(\mathbf{r},t)\,\rho_1(0,0)\rangle$ to $r \gg \xi$ since ξ here plays the same role of the mean free path r_0 far from the critical point. Thus, whenever the relation is not fulfilled, the problem of the temporal behavior of fluctuations relative to scattering has to be faced by resorting to new theoretical methods (see, for example, Kawasaki, 1971).

Besides the dynamical properties of the scattering system, the macroscopic single-scattering approach also becomes questionable near the critical point. As a matter of fact, while, as we have seen, the macroscopic result represented by the Einstein–Smolouchosky formula can be justified by means of a rigorous microscopic multiple-scattering treatment, the analogous formula for the critical region, which is obtained by using Eq. (3-93), has not received a microscopic justification. (An interesting calculation of the extinction coefficient has recently been given by Bedeaux and Mazur, 1973.) It seems unlikely that the formal simplicity of the macroscopic approach can be preserved very near the critical point (Bullough, 1967), since the extinction coefficient h markedly increases (*critical opalescence*) and one expects multiple-scattering contributions to become overwhelmingly important.

The current scattering experiments performed in the critical region in order to test the validity of the so-called *scaling laws* (see, for example, Swinney, 1974) concern situations far enough from the critical point to allow for an interpretation in terms of the macroscopic scattering theory as well as of the hydrodynamics linearized equations. Thus by inserting Eqs. (3-93) and (3-94) into Eq (3-23), one obtains

$$I(\eta) = \frac{V_{sc} I_0}{(4\pi r)^2} k_0^4 \sin^2 \gamma \left(\frac{d\varepsilon}{d\rho}\right)^2 \frac{2K_B T}{w_2} \frac{1}{\xi^2 + k_1^2} \qquad (3\text{-}97)$$

which has recently been the subject of direct experimental verification (Giglio and Benedek, 1969). Along the same lines, the decreasing of the spectral width of the Rayleigh peak, according to Eq. (3-91) and the divergence of c_p when approaching the critical point (Landau and Lifshitz, 1958, Chapter VIII) has been observed by many authors (see Swinney, 1974).

Very recently, an explicit microscopic evaluation of $I(\eta)$ near the critical point has been performed, which includes double scattering contributions (Oxtoby and Gelbart, 1974b) and implies a slight modification of the angular dependence of $I(\eta)$ as given by Eq. (3-97).

IV

Statistical Description of Electromagnetic Radiation

IV.1 General Remarks

The nature of electromagnetic radiation in the optical range of frequencies requires a statistical description. This is due to the intrinsically chaotic characteristics of the emission process, which does not allow us to make any exact prediction on the generated radiation. While this is the case for most available sources, a remarkable exception is furnished by laser sources. In any event, the random nature of the molecular scattering media introduces a statistical uncertainty on the scattered field, also in the case of a well-prescribed incident wave. Therefore, in this chapter we shall outline the main features of the statistical description of electromagnetic radiation.

Let us consider a general random process described in terms of a field variable whose single realization is indicated by $h(\mathbf{r}, t)$.

A statistical description of the physical process is contained in the set of nth-order *correlation functions* defined through the ensemble average

$$\langle h(\mathbf{r}_1, t_1) h(\mathbf{r}_2, t_2) \cdots h(\mathbf{r}_n, t_n) \rangle \tag{4-1}$$

for every value of n. As a matter of fact, the single realization does not possess, in general, a relevant meaning, and only averages of the kind in Eq. (4-1) are connected with measurable quantities. The information furnished by all possible ensemble averages is contained in the hierarchy of joint n-fold probability distributions

$$p_n(h_1, h_2, \ldots, h_n; x_1, x_2, \ldots, x_n), \qquad x_i \equiv (\mathbf{r}_i, t_i) \tag{4-2}$$

such that $p_n \, dh_1 \cdots dh_n$ represents the probability of finding a value between h_i and $h_i + dh_i$ for the field variable h_i at the space–time point x_i $(i = 1, \ldots, n)$. One has, in particular,

$$\int p_n(h_1, \ldots, h_n; x_1, \ldots, x_n) \, dh_1 \cdots dh_n = 1 \tag{4-3}$$

and

$$\langle h_1(x_1) \cdots h_n(x_n) \rangle = \int p_n(h_1, \ldots, h_n; x_1, \ldots, x_n) h_1 \cdots h_n \, dh_1 \cdots dh_n \tag{4-4}$$

According to the above considerations, the complete statistical description of the stochastic electric field $\mathbf{E}(\mathbf{r}, t)$ would require knowledge of the corresponding hierarchy of distribution functions. Also, for a large class of experiments, one can be satisfied with the less complete knowledge furnished by the associated correlation functions. We have already seen in Section I.5 how an electric field $\mathbf{E}(\mathbf{r}, t)$ can be represented by the complex analytic signal $\hat{\mathbf{E}}(\mathbf{r}, t)$ according to the relation

$$\mathbf{E}(\mathbf{r}, t) = \frac{\hat{\mathbf{E}}(\mathbf{r}, t) + \hat{\mathbf{E}}^*(\mathbf{r}, t)}{2} \tag{4-5}$$

where $E(r, t)$ is the positive frequency part of the electric field, which contains only terms varying as $e^{-i\omega t}$. Because of the formal mathematical convenience of considering the analytic signal, one can, in order to characterize the statistical properties of the radiation field, introduce a corresponding hierarchy of probability distributions. More precisely, the quantity

$$p_n(\hat{E}_1, ..., \hat{E}_n; x_1, ..., x_n) \, d^2 \hat{E}_1 \cdots d^2 \hat{E}_n \tag{4-6}$$

where

$$d^2 \hat{E}_K \equiv d(\mathrm{Re} \, \hat{E}_K) \, d(\mathrm{Im} \, \hat{E}_K), \qquad K = 1, 2, ..., n \tag{4-7}$$

represents the probability of finding a value between

$$\hat{E}_i \qquad \text{and} \qquad \hat{E}_i + d\hat{E}_i \, (d\hat{E}_i \equiv d(\mathrm{Re} \, \hat{E}_i) + i d(\mathrm{Im} \, \hat{E}_i))$$

for the field variable $\hat{E}$ at space–time point $x_i (i = 1, ..., n)$. These probability distributions allow us to evaluate directly any complicated correlation functions by means of the relation

$$\langle \hat{E}^*(x_1) \cdots \hat{E}^*(x_m) \, \hat{E}(x_{m+1}) \cdots \hat{E}(x_n) \rangle$$

$$= \int \hat{E}_1^* \cdots \hat{E}_m^* \hat{E}_{m+1} \cdots \hat{E}_n \, p_n(\hat{E}_1, ..., \hat{E}_n; x_1, ..., x_n)$$

$$\cdot \, d^2 \hat{E}_1 \cdots d^2 \hat{E}_n \tag{4-8}$$

We have already met a physical situation in which the second-order ensemble average $\langle \hat{E}^*(r, t) \cdot \hat{E}(r, t+\tau) \rangle$ was related to a significant quantity, namely, the power spectrum of a stationary electromagnetic field (Wiener–Khintchine theorem). Furthermore, second-order averages $\langle \hat{E}^*(r_1, t_1) \cdot \hat{E}(r_2, t_2) \rangle$, evaluated at different space–time points, have been shown to be completely sufficient in characterizing any kind of conventional interference experiments in optics. As a matter of fact, all of them deal essentially with an intensity measurement relative to a field resulting from the superposition of the fields coming from two or more space–time points:

$$\hat{\mathbf{E}}(\mathbf{r}, t) = \sum c_i \hat{\mathbf{E}}(\mathbf{r}_i, t_i) \tag{4-9}$$

where the c_i's are complex coefficients depending on the experimental geometry. By remembering the expression for the intensity in terms of the analytic signal as given in Section II.6, one has, for stationary situations,

$$I(\mathbf{r}, t) = \frac{c}{8\pi} \langle \hat{\mathbf{E}}^*(\mathbf{r}, t) \cdot \hat{\mathbf{E}}(\mathbf{r}, t) \rangle$$

$$= \frac{c}{8\pi} \sum_i \sum_j c_i c_j^* \langle \hat{\mathbf{E}}^*(\mathbf{r}_j, t_j) \cdot \hat{\mathbf{E}}(\mathbf{r}_i, t_i) \rangle \tag{4-10}$$

so that the quantities

$$\langle \hat{\mathbf{E}}^*(\mathbf{r}_j, t_j) \cdot \hat{\mathbf{E}}(\mathbf{r}_i, t_i) \rangle \tag{4-11}$$

completely determine the results of the experiment (Born and Wolf, 1970, Chapter X).

IV.2 The Existence of Higher-Order Correlation Effects: A Brief Historical Survey

The necessity of considering fourth-order ensemble averages such as $\langle |\hat{\mathbf{E}}(\mathbf{r}, t)|^2 |\hat{\mathbf{E}}(\mathbf{r}, t+\tau)|^2 \rangle$ received its first experimental evidence from the classical measurement of intensity correlations performed by Hanbury Brown and Twiss (1956), who introduced a new type of interferometer, which allows us to avoid the main difficulties associated with the use of the Michelson stellar interferometer. In fact, by means of the latter apparatus one performs a conventional interference experiment leading to the measurement of the second-order average $\langle \hat{\mathbf{E}}^*(\mathbf{r}_1, t) \cdot \hat{\mathbf{E}}(\mathbf{r}_2, t+\tau) \rangle$, a quantity that depends on the relative phases of the two signals, which are combined before detection. These phases are sensitive to spurious random fluctuations associated with the structure of the interferometer and the variations of the index of refraction along the optical path, a fact that can limit the applicability of the method. In the Hanbury Brown and Twiss experiment (see Fig. 4.1) this difficulty is avoided by detecting the intensities of two signals impinging on the two mirrors M_1 and M_2, so that all phase fluctuations are eliminated, and then by transmitting the two low-frequency signals to a multiplier device by means of a suitable delay line. The measurement reveals the existence of an intensity correlation whose behavior as a function of the distance between the two mirrors furnishes information on the angular diameter of the stellar source.

The result of this experiment, besides its practical relevance, is important for its implications if one considers the basic description of electromagnetic radiation both from a classical and a quantum-mechanical point of view (Purcell, 1956; Hanbury

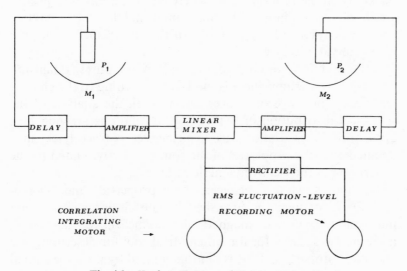

Fig. 4.1 *Hanbury Brown and Twiss experiment.*

Brown and Twiss, 1957). Classically, the experiment is consistent with the wave description of light once the presence of intensity fluctuations due to a source consisting of many radiators emitting at random is taken into account.

The quantum-mechanical interpretation is more refined in that it relies on both the nature of the emission process and the corpuscular behavior of photons. In fact, if the photons were independent particles, their number would be Poisson distributed in the case of a completely random source. This would imply the absence of fluctuations, and thus of correlations, in the detected intensities (see Sections IV.7 and IV.8). In effect, the photons behave as indistinguishable particles, such that only states symmetrical between them can occur in nature. This implies a tendency toward bunching, which accounts for the presence of a correlation in the Hanbury Brown and Twiss experiment. If one associates a wave packet with each photon, it is clear that photon bunching can occur only because of partial overlapping between different packets. Since the temporal duration of each packet is on the order of $1/\Delta v$, where Δv represents the spectral width of the radiation, the indistinguishability between two photons plays a relevant role only if they are detected within a time interval $T \leqslant 1/\Delta v$.

If the source is not completely random, the purely quantum-mechanical bunching effect is modified according to the characteristics of the emission process. In general, the analysis of the statistical distribution of the number of photons arriving in a given time interval allows us, in principle, to trace back, in a detailed way, the properties of the source directly related to the generation of the electromagnetic radiation.

The analytical link between these properties and *photon-counting* experiments is furnished by the all-order correlation functions of the electromagnetic field, which have been recognized as the most natural mathematical tool for describing the statistical properties of the radiation emitted by unconventional sources (Wolf, 1963; Glauber, 1963a). As a basic example, the fundamental differences between a laser beam and a beam emitted by a conventional source only emerge when considering correlations of order higher than second. We shall see how these correlations are connected with photon-counting measurements, a kind of experiment that is feasible at optical frequencies, where the corpuscular character of radiation can be made apparent (see, for example, Klauder and Sudarshan, 1968, and Peřina, 1971). We observe here that an ideal laser constitutes a peculiar source in that the photon bunching effect is completely canceled by the strong correlation between the microfields emitted by the various atoms. In effect, the photon number distribution turns out to be Poisson and, equivalently, the intensity is a well-prescribed quantity, as it would be for a beam of independent classical particles emitted by a random source.

The statistical description of the electromagnetic radiation we have dealt with up to now is in terms of classical quantities. On the other hand, we have just observed that the treatment of a phenomenon in which the individuality of a single photon is detected cannot be purely classical. A correct approach requires then the quantization of both the electromagnetic field and the detecting atomic system. Quantum theory allows us to express the photon-counting distribution in terms of quantum-mechanical correlation functions, which are the natural generalization of the classical ones introduced in Section IV.1, whose use is "a priori" justified only in connection with the application of the correspondence principle, that is, when dealing with strong electromagnetic fields. There exists, however, a rather general representation of the electromagnetic field (Glauber, 1963b; Sudarshan, 1963), whose most relevant feature is allowing us to perform all calculations in formal analogy with the semi-classical treatment where the electromagnetic field is not quantized, independently of the strength of the field.

The generality of this quantum-mechanical representation has for a long time been the object of discussion. It seems reasonable to say that, if there is a sense in which it always exists, its usefulness is limited to situations in which it assumes manageable analytical form (Glauber, 1967). As we shall see, there are relevant examples of light fields for which this is just the case.

IV.3 Quantization of the Electromagnetic Field

The starting point of the standard technique of quantization of the electromagnetic radiation in vacuo is Maxwell's equations written in the absence of charges and currents. In this case, the electric field $\mathbf{E}(\mathbf{r}, t)$, and the magnetic field $\mathbf{H}(\mathbf{r}, t)$ can be derived in terms of the single vector potential $\mathbf{A}(\mathbf{r}, t)$ (Born and Wolf, 1970, Chapter II), which obeys the wave equation

$$\nabla^2 \mathbf{A}(\mathbf{r}, t) - \frac{1}{c^2} \frac{\partial^2 \mathbf{A}(\mathbf{r}, t)}{\partial t^2} = 0 \tag{4-12}$$

together with the transversality condition

$$\nabla \cdot \mathbf{A}(\mathbf{r}, t) = 0 \tag{4-13}$$

in the form

$$\mathbf{E}(\mathbf{r}, t) = -\frac{1}{c} \frac{\partial \mathbf{A}(\mathbf{r}, t)}{\partial t} \tag{4-14}$$

$$\mathbf{H}(\mathbf{r}, t) = \nabla \times \mathbf{A}(\mathbf{r}, t) \tag{4-15}$$

Let us suppose that the field is confined in a cavity with perfectly conducting walls. It is convenient to look for a general solution of Eqs. (4-12) and (4-13), with the associated boundary conditions, as an expansion with real coefficients $q_l(t)$ in terms of a complete set of real orthogonal mode functions $\mathbf{u}_l(\mathbf{r})$ (see, for example, Louisell, 1964, Chapter IV):

$$\mathbf{A}(\mathbf{r}, t) = \sum_l q_l(t) \mathbf{u}_l(\mathbf{r}) \tag{4-16}$$

Substitution of Eq. (4-16) into Eqs. (4-12) and (4-13) gives, respectively,

$$\nabla^2 \mathbf{u}_l(\mathbf{r}) + \frac{\omega_l^2}{c^2} \mathbf{u}_l(\mathbf{r}) = 0 \tag{4-17}$$

$$\frac{d^2 q_l(t)}{dt^2} + \omega_l^2 q_l(t) = 0 \tag{4-18}$$

ω_l^2 being the separation constant, and

$$\nabla \cdot \mathbf{u}_l(\mathbf{r}) = 0 \tag{4-19}$$

The fact that the $\mathbf{u}_l(\mathbf{r})$'s must satisfy the boundary conditions

$$\begin{aligned} \mathbf{u}_l \times \mathbf{n} &= 0 \\ \nabla \times \mathbf{u}_l \cdot \mathbf{n} &= 0 \end{aligned} \tag{4-20}$$

$\mathbf{n}$ being the normal to the cavity walls, ensures that Eq. (4-17) admits a denumerable set of orthogonal eigenfunctions $\mathbf{u}_l(\mathbf{r})$ with corresponding eigenvalues ω_l^2/c^2.

The total energy of the field

$$H_0 = \frac{1}{8\pi} \int_{\text{cavity}} (\mathbf{E}^2 + \mathbf{H}^2) \, d\mathbf{r} \tag{4-21}$$

can now be easily evaluated by inserting Eq. (4-16) into Eqs. (4-14) and (4-15), thus obtaining

$$H_0 = \frac{1}{8\pi c^2} \int_{\text{cavity}} \sum_l \sum_m [\dot{q}_l(t) \dot{q}_m(t) \mathbf{u}_l(\mathbf{r}) \cdot \mathbf{u}_m(\mathbf{r})$$
$$+ c^2 q_l(t) q_m(t) \nabla \times \mathbf{u}_l(\mathbf{r}) \cdot \nabla \times \mathbf{u}_m(\mathbf{r})] \, d\mathbf{r} \tag{4-22}$$

By using the vector identity now,

$$\nabla \times \mathbf{u}_l \cdot \nabla \times \mathbf{u}_m = \mathbf{u}_m \cdot [\nabla \times (\nabla \times \mathbf{u}_l)] + \nabla \cdot [\mathbf{u}_m \times (\nabla \times \mathbf{u}_l)] \tag{4-23}$$

and the fact that, according to Eqs. (4-17) and (4-19),

$$\nabla \times (\nabla \times \mathbf{u}_l) = \nabla(\nabla \cdot \mathbf{u}_l) - \nabla^2 \mathbf{u}_l = \frac{\omega_l^2}{c^2} \mathbf{u}_l \tag{4-24}$$

Eq. (4-22) can be written as

$$H_0 = \tfrac{1}{2}\sum_l [\dot{q}_l^2(t) + \omega_l^2 q_l^2(t)]$$

$$+ \frac{1}{8\pi} \sum_l \sum_m q_l(t) q_m(t) \int_{\text{cavity}} \nabla \cdot [\mathbf{u}_m \times (\nabla \times \mathbf{u}_l)]\, d\mathbf{r} \quad (4\text{-}25)$$

where the orthonormality condition

$$\int_{\text{cavity}} \mathbf{u}_l \cdot \mathbf{u}_m \, d\mathbf{r} = 4\pi c^2 \delta_{lm} \quad (4\text{-}26)$$

has been introduced. If one observes that the last term on the right-hand side of Eq. (4-25) vanishes, as can be easily seen by applying Gauss's theorem and remembering Eq. (4-20), we finally obtain

$$H_0 = \tfrac{1}{2}\sum_l [p_l^2(t) + \omega_l^2 q_l^2(t)] \quad (4\text{-}27)$$

where

$$p_l(t) = \frac{dq_l}{dt} \quad (4\text{-}28)$$

In this way, the electromagnetic field is described in terms of a set of independent couples of conjugate variables q_l and p_l relative to a set of independent harmonic oscillators. The quantization of the electromagnetic field is now achieved by regarding the q_l's and p_l's as Hermitian operators obeying the commutation relations

$$[p_l, p_m] = [q_l, q_m] = 0$$
$$[q_l, p_m] = i\hbar \delta_{lm} \quad (4\text{-}29)$$

according to a basic postulate of quantum mechanics. The q_l's and p_l's will be explicitly time independent or not according to whether or not one uses the Schrödinger picture. The standard procedure of quantization of the harmonic oscillator consists then in the introduction of a pair of non-Hermitian operators

a_l^+ and a_l by means of the equations

$$q_l = \left(\frac{\hbar}{2\omega_l}\right)^{1/2} [a_l^+ + a_l] \qquad (4\text{-}30)$$

$$p_l = i\left(\frac{\hbar\omega_l}{2}\right)^{1/2} [a_l^+ - a_l] \qquad (4\text{-}31)$$

It is easily seen from Eqs. (4-30) and (4-31) that a_l and a_l^+ are Hermitian conjugate operators, while Eq. (4-29) shows that they obey the commutation relations

$$\begin{aligned} [a_l, a_m^+] &= \delta_{lm} \\ [a_l, a_m] &= [a_l^+, a_m^+] = 0 \end{aligned} \qquad (4\text{-}32)$$

The Hamiltonian of the system immediately follows from Eqs. (4-27) and (4-30)–(4-32) as

$$H_0 = \sum_l H_l = \sum_l \hbar\omega_l(a_l^+ a_l + \tfrac{1}{2}) \qquad (4\text{-}33)$$

If one now chooses to use the Heisenberg picture, which is the most appropriate if a comparison between classical and quantum treatment has to be made, the time evolution of $a_l(t)$ and $a_l^+ (t)$ is determined by the Heisenberg equations of motion

$$i\hbar\frac{da_l(t)}{dt} = [a_l(t), H_0] = \hbar\omega_l a_l(t) \qquad (4\text{-}34)$$

$$i\hbar\frac{da_l^+ (t)}{dt} = [a_l^+ (t), H_0] = -\hbar\omega_l a_l^+ (t) \qquad (4\text{-}35)$$

having taken into account Eqs. (4-32) and (4-33). One has, according to Eqs. (4-34) and (4-35),

$$\begin{aligned} a_l(t) &= a_l e^{-i\omega_l t} \\ a_l^+ (t) &= a_l^+ e^{i\omega_l t} \end{aligned} \qquad (4\text{-}36)$$

where $a_l = a_l(0)$ and $a_l^+ = a_l^+ (0)$ are henceforth these operators in the Schrödinger picture.

As is well known (see, for example, Louisell, 1964, Chapter II), the eigenvalues n_l of the operator associated with the eigenvalue equation

$$a_l^+ a_l |n_l\rangle = n_l |n_l\rangle \qquad (4\text{-}37)$$

are furnished by all the nonnegative integers, so that n_l can be interpreted as the number of energy quanta in the mode l.

This allows us to give $a_l^+ a_l$ the meaning of number operator. Furthermore, a_l^+ and a_l are usually termed *creation* and *annihilation* operators since they can be shown, respectively, to increase and lower by one the number of quanta when operating on the eigenstates $|n_l\rangle$ of $a_l^+ a_l$ according to

$$a_l^+ |n_l\rangle = (n_l + 1)^{1/2} |n_l + 1\rangle \qquad (4\text{-}38)$$

$$a_l |n_l\rangle = (n_l)^{1/2} |n_l - 1\rangle \qquad (4\text{-}39)$$

The preceding considerations summarize the properties of the vector potential operator, which, according to Eqs. (4-16) and (4-30), can be written in the Heisenberg picture as

$$\mathbf{A}(\mathbf{r}, t) = \sum_l \left(\frac{\hbar}{2\omega_l}\right)^{1/2} [a_l e^{-i\omega_l t} + a_l^+ e^{i\omega_l t}] \mathbf{u}_l(\mathbf{r}) \qquad (4\text{-}40)$$

The electromagnetic radiation field is then completely specified from a quantum-mechanical point of view once the initial state of the system is assigned in terms of the $|n_l\rangle$'s.

Let us now turn our attention to the physical meaning of a single mode l of the electromagnetic field. Assuming we are dealing with a cubical cavity of volume L^3 and sides lying on the three axes (x, y, z) of an orthogonal reference frame with respective unit vectors $\mathbf{i}, \mathbf{j}, \mathbf{h}$, Eqs. (4-19) and (4-20) are satisfied by

$$
\begin{aligned}
\mathbf{u}_l(\mathbf{r}) = \mathbf{u}_{l_1, l_2, l_3}(\mathbf{r}) = {} & \alpha \cos(k_{l_1} x) \sin(k_{l_2} y) \sin(k_{l_3} z) \mathbf{i} \\
& + \beta \sin(k_{l_1} x) \cos(k_{l_2} y) \sin(k_{l_3} z) \mathbf{j} \\
& + \gamma \sin(k_{l_1} x) \sin(k_{l_2} y) \cos(k_{l_3} z) \mathbf{h} \qquad (4\text{-}41)
\end{aligned}
$$

if

$$\alpha k_{l_1} + \beta k_{l_2} + \gamma k_{l_3} = 0 \qquad (4\text{-}42)$$

and

$$k_{l_i} = \frac{2\pi l_i}{L}, \qquad i = 1, 2, 3 \qquad (4\text{-}43)$$

with l_i an integer from $-\infty$ to $+\infty$. Furthermore, Eqs. (4-17) and (4-20) ensure that ω_l assumes the discrete values

$$\omega_l = c(k_{l_1}^2 + k_{l_2}^2 + k_{l_3}^2)^{1/2} \qquad (4\text{-}44)$$

One has to observe that the possibility of expanding the electromagnetic field in standing waves obviously implies the vanishing of the total momentum of the field in the cavity

$$\mathbf{G} = \frac{1}{4\pi c} \int_{\text{cavity}} (\mathbf{E} \times \mathbf{H}) \, d\mathbf{r} = 0 \qquad (4\text{-}45)$$

Since the cavity momentum pertaining to each mode may be seen to vanish as well, a_l^+ and a_l can be interpreted as creation and annihilation operators relative to quanta with energy $\hbar\omega_l$ with zero momentum, a fact that does not allow us to interpret the energy quanta as photons.

It is possible to consider a more general electromagnetic radiation field than the one described by means of standing waves, by introducing a representation of $\mathbf{A}(\mathbf{r}, t)$ based on the use of *traveling plane waves* (Louisell, 1964, Chapter IV). More precisely, it can be shown that the general solution of Eq. (4-12) subjected to the transversality condition and to the periodic boundary conditions

$$\mathbf{A}(\mathbf{r} + L\mathbf{i}, t) = \mathbf{A}(\mathbf{r} + L\mathbf{j}, t) = \mathbf{A}(\mathbf{r} + L\mathbf{h}, t) = \mathbf{A}(\mathbf{r}, t) \quad (4\text{-}46)$$

can be written in the form of an expansion with complex coefficients:

$$\mathbf{A}(\mathbf{r}, t) = \sum_l \sum_{\sigma=1}^{2} \mathbf{e}_{l\sigma} \{ c_{l\sigma} \exp[i(\mathbf{k}_l \cdot \mathbf{r} - \omega_l t)] $$
$$+ c_{l\sigma}^* \exp[-i(\mathbf{k}_l \cdot \mathbf{r} - \omega_l t)] \} \qquad (4\text{-}47)$$

with
$$\omega_l = ck_l \tag{4-48}$$
and

$$\mathbf{k}_l \equiv \mathbf{k}_{l_1, l_2, l_3} = \frac{2\pi}{L}(l_1\,\mathbf{i} + l_2\,\mathbf{j} + l_3\,\mathbf{h}) \tag{4-49}$$

where l_i ($i = 1, 2, 3$) is an integer ranging from $-\infty$ to $+\infty$ and the $\mathbf{e}_{l_\sigma}$'s are two unit vectors specifying the polarization of each *mode* (l, σ), which can be chosen to be mutually orthogonal and must satisfy the relation

$$\mathbf{e}_{l_\sigma} \cdot \mathbf{k}_l = 0 \tag{4-50}$$

While the potential vector given in Eq. (4-16) is always expressible in the form of Eq. (4-47), as one can easily see by using Eqs. (4-41)–(4-44), it is not always possible to write $\mathbf{A}(\mathbf{r}, t)$ as in Eq. (4-16) starting from Eq. (4-47). This derives from the boundary conditions valid for the radiation contained in a cavity with perfectly conducting walls [see Eq. (4-20)], which are more restrictive than the simple periodic condition of Eq. (4-46). The quantization of the vector potential expressed in terms of traveling waves can be performed, following a method analogous to the one yielding Eq. (4-40) (Louisell, 1964, Chapter IV). The final expression for the vector potential operator is

$$\mathbf{A}(\mathbf{r}, t) = \frac{c}{L^{3/2}} \sum_{l, \sigma} \left(\frac{h}{\omega_l}\right)^{1/2} \mathbf{e}_{l_\sigma} \{ a_{l_\sigma} \exp[i(\mathbf{k}_l \cdot \mathbf{r} - \omega_l t)]$$
$$+ a_{l_\sigma}^+ \exp[-i(\mathbf{k}_l \cdot \mathbf{r} - \omega_l t)] \} \tag{4-51}$$

from which the following relations can be deduced:

$$H_0 = \sum_{l, \sigma} \hbar\omega_l(a_{l_\sigma}^+ a_{l_\sigma} + \tfrac{1}{2}) \tag{4-52}$$

$$\mathbf{G} = \sum_{l, \sigma} \hbar\mathbf{k}_l a_{l_\sigma}^+ a_{l_\sigma} \tag{4-53}$$

with the help of Eqs. (4-14), (4-15), (4-21), and (4-45).

The Hermitian conjugate operators $a_{l\sigma}$ and $a_{l\sigma}^+$ satisfy the commutation relations

$$
\begin{aligned}
[a_{l\sigma}, a_{l'\sigma'}^+] &= \delta_{ll'}\delta_{\sigma\sigma'} \\
[a_{l\sigma}, a_{l'\sigma'}] &= [a_{l\sigma}^+, a_{l'\sigma'}^+] = 0
\end{aligned}
\tag{4-54}
$$

and can be interpreted as annihilation and creation operators of a photon of energy $\hbar\omega_l$, momentum $\mathbf{k}_l$, and polarization σ, since obvious generalizations of Eqs. (4-37)–(4-39) exist.

Finally, it is hardly necessary to note that the total momentum $\mathbf{G}$ of the electromagnetic field whose vector potential $\mathbf{A}(\mathbf{r}, t)$ is given by Eq. (4-51), or classically by Eq. (4-47), is in general different from zero.

IV.4 The Photon Detection Process

Classical optics has always tacitly assumed the existence of a device capable of measuring the intensity of the electromagnetic field (averaged in practice over many periods of oscillation of the field due to the finite resolution time of the instrument). In effect, the detection process is an intrinsically quantum-mechanical one since it is related to the individual nature of the single photon, and we shall present here the correct procedure by which both the radiation field and the detecting system are quantized (Glauber, 1965). Although a number of mechanisms capable of revealing photons are, in principle, conceivable (stimulated emission, Compton effect, and so on), we shall deal here with the most realistic one, based on photon absorption. The simplest counting apparatus of this kind can be schematized by a single atom free to undergo photoelectric effect, after which the ejected electron is observed.

The system constituted by the atom and the field is described in terms of a Hamiltonian composed of a free part H_0 and an interaction part H_1, which, in the dipole approximation, i.e., for wavelengths much greater than the atomic dimension, is (see Section I.3)

$$H_1 = -e \sum_{\gamma} \mathbf{q}_\gamma(t) \cdot \mathbf{E}(\mathbf{r}, t) \tag{4-55}$$

where $\mathbf{q}_\gamma(t)$ represents the position operator of the γth electron relative to the nucleus, whose position is indicated by $\mathbf{r}$, and $\mathbf{E}(\mathbf{r}, t)$ is the electric field operator.

In the interaction representation, assuming that H_0 is the sum of the Hamiltonians of the atom H_{01} and of the free field H_{02} (which amounts to saying that the detector is far from the electromagnetic sources), the temporal behavior of the $\mathbf{q}_\gamma$'s and of $\mathbf{E}(\mathbf{r}, t)$ turns out to be

$$\begin{aligned}
\mathbf{q}_\gamma(t) &= \exp\left(\frac{i}{\hbar} H_0 \, t\right) \mathbf{q}_\gamma(0) \exp\left(-\frac{i}{\hbar} H_0 \, t\right) \\
&= \exp\left(\frac{i}{\hbar} H_{01} \, t\right) \mathbf{q}_\gamma(0) \exp\left(-\frac{i}{\hbar} H_{01} \, t\right) \tag{4-56}
\end{aligned}$$

$$\begin{aligned}
\mathbf{E}(\mathbf{r}, t) &= \exp\left(\frac{i}{\hbar} H_0 \, t\right) \mathbf{E}(\mathbf{r}, 0) \exp\left(-\frac{i}{\hbar} H_0 \, t\right) \\
&= \exp\left(\frac{i}{\hbar} H_{02} \, t\right) \mathbf{E}(\mathbf{r}, 0) \exp\left(-\frac{i}{\hbar} H_{02} \, t\right) \tag{4-57}
\end{aligned}$$

that is, the behavior of the free operators in the Heisenberg picture.

Let the system be at time t_0 in the state $|g\rangle|i\rangle$, where $|g\rangle$ is the ground state of the atom[†] and $|i\rangle$ the state of the field. Assuming that the interaction is present in the interval (t_0, t), its total effect consists of modifying the state of the system in this interval according to

$$|t\rangle = \left[1 - \frac{i}{\hbar} \int_{t_0}^{t} H_1(t')\, dt'\right] |g\rangle|i\rangle \tag{4-58}$$

[†] In effect, the ground state of the atom is the initial state only if thermal excitation is negligible.

to the first order in the perturbation. By considering a state of the unperturbed system $|a\rangle|f\rangle$ in the Heisenberg picture, where $|a\rangle$ is an energy eigenstate of the atomic system distinct from $|g\rangle$ and $|f\rangle$ a state of the field, the transition probability from the state $|g\rangle|i\rangle$ to $|a\rangle|f\rangle$ is the square modulus of the amplitude

$$\langle a|\langle f|t\rangle = -\frac{ie}{\hbar}\sum_{\gamma}\int_{t_0}^{t}\langle a|\,\mathbf{q}_{\gamma}(t')|g\rangle \cdot \langle f|\,\mathbf{E}(\mathbf{r},t')|i\rangle\,dt' \qquad (4\text{-}59)$$

This expression can be rewritten in turn as

$$\langle a|\langle f|t\rangle = -\frac{ie}{\hbar}\int_{t_0}^{t}\exp(i\omega_{ag}\,t')\,\mathbf{M}_{ag} \cdot \langle f|\,\mathbf{E}(\mathbf{r},t')|i\rangle\,dt' \qquad (4\text{-}60)$$

where

$$\mathbf{M}_{ag} = \sum_{\gamma}\langle a|\,\mathbf{q}_{\gamma}(0)|g\rangle \qquad (4\text{-}61)$$

and $\hbar\omega_{ag} = E_a - E_g$, E_a and E_g representing the energy eigenvalues relative to the atomic state $|a\rangle$ and $|g\rangle$. One can now observe that according to Eqs. (4-14) and (4-51) one has

$$\mathbf{E}(\mathbf{r},t) = i\sum_{l,\sigma}\left(\frac{\hbar\omega_l}{L^3}\right)^{1/2}\mathbf{e}_{l\sigma}\{a_{l\sigma}\exp[i(\mathbf{k}_l\cdot\mathbf{r}-\omega_l t)]$$
$$- a_{l\sigma}^{+}\exp[-i(\mathbf{k}_l\cdot\mathbf{r}-\omega_l t)]\} \qquad (4\text{-}62)$$

which can be rewritten as the sum of the two Hermitian conjugate operators $\mathbf{E}^{(+)}(\mathbf{r},t)$ and $\mathbf{E}^{(-)}(\mathbf{r},t)$, the so-called *positive* and *negative* frequency part of the electric field

$$\mathbf{E}(\mathbf{r},t) = \mathbf{E}^{(+)}(\mathbf{r},t) + \mathbf{E}^{(-)}(\mathbf{r},t) \qquad (4\text{-}63)$$

where

$$\mathbf{E}^{(+)}(\mathbf{r},t) = i\sum_{l,\sigma}\left(\frac{\hbar\omega_l}{L^3}\right)^{1/2}\mathbf{e}_{l\sigma}\,a_{l\sigma}\exp[i(\mathbf{k}_l\cdot\mathbf{r}-\omega_l t)] \qquad (4\text{-}64)$$

The operators $\mathbf{E}^{(+)}$ and $\mathbf{E}^{(-)}$ contain, respectively, only annihilation and creation operators and time exponentials of the form $\exp(-i\omega t)$ and $\exp(i\omega t)$.

Substitution of Eq. (4-63) into Eq. (4-60) immediately shows that the only nonnegligible contribution to the transition amplitude can come from $\mathbf{E}^{(+)}(\mathbf{r}, t)$, whenever the relation $t - t_0 \gg 1/\omega_{ag}$ is satisfied, as always occurs in practice. This fact can readily be explained in terms of the uncertainty principle, according to which the energy can be conserved with an accuracy $\Delta E = \hbar/(t - t_0)$. On the other hand, the creation of a photon associated with $\mathbf{E}^{(-)}(\mathbf{r}, t)$ yields $\Delta E \geqslant \hbar\omega_{ag} \gg \hbar/(t - t_0)$, a relation that makes this process very unlikely. According to these considerations, Eq. (4-60) can be rewritten as

$$\langle a| \langle f| t \rangle = -\frac{ie}{\hbar} \int_{t_0}^{t} \exp(i\omega_{ag} t') \mathbf{M}_{ag} \cdot \langle f| \mathbf{E}^{(+)}(\mathbf{r}, t') |i \rangle \, dt' \quad (4\text{-}65)$$

The total transition probability $P_{g \to a}$ from the initial atomic state $|g\rangle$ to the final state $|a\rangle$ is now readily obtained by summing the contribution pertaining to each final state $|f\rangle$ of the field and taking into account the statistical uncertainty usually present in the electric field by ensemble averaging over the initial states $|i\rangle$. One obtains[†]

$$P_{g \to a} = \left\{ \sum_{|f\rangle} |\langle a| \langle f| t \rangle|^2 \right\}_{\text{average over } |i\rangle}$$

$$= \left(\frac{e}{\hbar} \right)^2 \int_{t_0}^{t} dt' \int_{t_0}^{t} dt'' \exp[i\omega_{ag}(t'' - t')] \mathbf{M}_{ag}^* \, \mathbf{M}_{ag} : \mathbf{G}^{(1)}(\mathbf{r}t', \mathbf{r}t'')$$
$$(4\text{-}66)$$

where the tensor $\mathbf{G}^{(1)}(\mathbf{r}t', \mathbf{r}t'')$ is the first-order, quantum-mechanical, correlation function of the electric field defined as

$$\mathbf{G}^{(1)}(\mathbf{r}t', \mathbf{r}t'') = \{\langle i| \mathbf{E}^{(-)}(\mathbf{r}, t') \mathbf{E}^{(+)}(\mathbf{r}, t'') |i\rangle\}_{\text{average over } |i\rangle} \quad (4\text{-}67)$$

first introduced by Glauber (1963a).

As for the detection process itself, one observes that a given atomic state $|a\rangle$ corresponds to the ejection of an electron (if

[†]The symbol $\mathbf{T}_1 : \mathbf{T}_2$ indicates the scalar product between the two tensors $\mathbf{T}_1$ and $\mathbf{T}_2$ defined as $\mathbf{T}_1 : \mathbf{T}_2 = \sum_{k,i}^{1,3} (T_1)_{ik} (T_2)_{ki}$.

any) in a well-defined state. This electron has a certain probability $R(a)$ to be revealed, which depends on the characteristics of the photoelectron detector employed as well as on the state $|a\rangle$. Thus the probability $p^{(1)}(t)$ of counting a photon is obtained by weighting Eq. (4-66) by means of the probability function $R(a)$. The net outcome of this procedure is

$$p^{(1)}(t) = \sum_{|a\rangle} R(a) P_{g \to a}(t) = \int_{t_0}^{t} dt' \int_{t_0}^{t} dt'' \, \mathbf{s}(t'' - t') : \mathbf{G}^{(1)}(rt', rt'')$$

$$(4\text{-}68)$$

where the tensor $\mathbf{s}(t)$ summarizes the response of the detecting atomic system and is given by the Fourier transform of the *sensitivity function*

$$\mathbf{s}(\omega) = 2\pi \left(\frac{e}{\hbar}\right)^2 \sum_a R(a) \, \mathbf{M}_{ag}^* \, \mathbf{M}_{ag} \, \delta(\omega - \omega_{ag}) \qquad (4\text{-}69)$$

If one considers a *broadband* detector, corresponding to a sensitivity independent of ω in the relevant frequency range, $\mathbf{s}(t)$ behaves as a δ-function[†] so that

$$p^{(1)}(t) = \mathbf{s} : \int_{t_0}^{t} \mathbf{G}^{(1)}(r\underline{t}', rt') \, dt' \qquad (4\text{-}70)$$

We note at this point that no essential limitation is introduced by considering the electric field and hence $\mathbf{G}$ as scalar quantities, which practically corresponds to using a suitable polarization filter placed in front of the detector. Accordingly, we shall refer hereafter to this simplified situation, so that $\mathbf{E}$, $\mathbf{G}^{(1)}$, and $\mathbf{s}$ will be replaced by the scalar quantities E, $G^{(1)}$, and s. The expression given by Eq. (4-70) is fully quantum mechanical, since both the radiation and the detecting system have been quantized. In the

---

[†] More precisely, $\mathbf{s}(t)$ behaves as a δ-function only if $\Delta\omega \gg \delta\omega$, $\Delta\omega \gg 1/(t - t_0)$, where $\Delta\omega$ represents a frequency range over which $\mathbf{s}(\omega)$ is nearly constant and $\delta\omega$ is the radiation bandwidth. The second relation sets a lower limit on the detection time (Glauber, 1965). On the other hand, only fluctuations taking place in a time interval greater than $t - t_0$ can actually be observed.

classical limit for the radiation field, one expects the quantum-mechanical operators $E^{(+)}$ and $E^{(-)}$ to be, respectively, replaced by the analytic signal $\hat{E}$ and its complex conjugate. As a matter of fact, this can be rigorously proved (Glauber, 1963b) and it is in perfect agreement with the semiclassical treatment in which only the atomic system is quantized (Mandel *et al.*, 1964). In this situation, the quantum-mechanical quantity $G^{(1)}$ can be approximated by the classical ensemble average

$$G^{(1)}(\mathbf{r}t, \mathbf{r}t) = \langle \hat{E}^*(\mathbf{r}, t)\,\hat{E}(\mathbf{r}, t)\rangle \qquad (4\text{-}71)$$

We now observe that, in a real counting device, $t - t_0$ contains a great number of oscillation periods of the radiation field, so that $p^{(1)}(t)$ is, in practice, an average of $G^{(1)}(\mathbf{r}t, \mathbf{r}t)$ over many periods. On the other hand, it can be easily shown (Born and Wolf, 1970, Chapter X) that $\hat{E}^*(\mathbf{r}, t)\,\hat{E}(\mathbf{r}, t)$ is, for quasi-mono-chromatic fields, proportional to the time average over few periods of the instantaneous intensity $I(\mathbf{r}, t)$, so that Eq. (4-70) is, in the classical limit, equivalent to

$$p^{(1)}(t) = s' \int_{t_0}^{t} \langle I(\mathbf{r}, t')\rangle\, dt' \qquad (4\text{-}72)$$

with $s' = (8\pi/c)\,s$.

In the case of an actual detector containing a great number N of independent identical atoms, which are supposed to be uniformly illuminated by the radiation field, the average number of counts $\langle C(t)\rangle$ recorded in the time interval $t - t_0$ is given by

$$\langle C(t)\rangle = Np^{(1)}(t) \qquad (4\text{-}73)$$

Equations (4-72) and (4-73) furnish a justification for the classical statement that the response of a photoelectric detector is proportional to the incident intensity. The quantum generalization given by Eq. (4-70) shows that the relevant operator corresponding to the classical intensity is $E^{(-)}(\mathbf{r}, t)\,E^{(+)}(\mathbf{r}, t)$. This is a significant result since it provides an example in which the quantization of a classical law is not a priori unique and requires a precise physical insight for determining the correct ordering of the involved operators.

IV.5   The *n*-Photon Detection Process

The basic analytical technique developed in the previous section and leading to Eq. (4-70) can be applied as well to the more general situation in which n identical one-atom detectors placed at different positions r_i ($i = 1, 2 ..., n$) are acted upon by the radiation field (which, for each atom, is assumed to be unaffected by the presence of the surrounding ones). One looks in this case for the probability $p^{(n)}(t)$ of counting n photons in a time interval $t - t_0$ (neglecting all contributions in which a single atom undergoes a photoionization process with more than one photon involved). Under the same assumptions leading to Eq. (4-70) it is possible to obtain the relation (Glauber, 1965)

$$p^{(n)}(t) = s^n \int_{t_0}^t dt_1 \cdots \int_{t_0}^t dt_n \, G^{(n)}(r_1 t_1, \ldots, r_n t_n, r_n t_n, \ldots, r_1 t_1) \quad (4\text{-}74)$$

where the nth order quantum-mechanical correlation function $G^{(n)}$ is defined as

$$G^{(n)}(r_1 t_1, \ldots, r_n t_n, r_n t_n, \ldots, r_1 t_1)$$

$$= \{\langle i| E^{(-)}(r_1, t_1) \cdots E^{(-)}(r_n, t_n)$$

$$\times E^{(+)}(r_n, t_n) \cdots E^{(+)}(r_1, t_1)|i\rangle\}_{\text{average over } |i\rangle} \quad (4\text{-}75)$$

Equation (4-75) is the direct generalization of Eq. (4-67) and represents the statistical and quantum expectation value of the well-ordered product of the negative and positive frequency parts of the electric field operator associated, respectively, with the creation and annihilation of photons. The classical limit of Eq. (4-75), that is, the generalization of Eq. (4-71), is obtained as before by substituting the positive and negative frequency part operators with the classical analytic signal and its complex conjugate, thus obtaining

$$G^{(n)}(\mathbf{r}_1 \, t_1, \dots, \mathbf{r}_n \, t_n, \mathbf{r}_n \, t_n, \dots, \mathbf{r}_1 \, t_1)$$
$$= \langle \hat{E}^*(\mathbf{r}_1, t_1) \cdots \hat{E}^*(\mathbf{r}_n, t_n) \, \hat{E}(\mathbf{r}_n, t_n) \cdots \hat{E}(\mathbf{r}_1, t_1) \rangle \quad (4\text{-}76)$$

This in turn implies the generalization of Eq. (4-72) in the form

$$p^{(n)}(t) = s'^n \int_{t_0}^t dt_1 \cdots \int_{t_0}^t dt_n \, \langle I(\mathbf{r}_1, t_1) \cdots I(\mathbf{r}_n, t_n) \rangle \quad (4\text{-}77)$$

The previous considerations show the possibility of performing experiments in which higher-order statistical properties of the radiation field can, in principle, be measured. In this respect, we remember that, as far as classical fields are considered, complete knowledge of the statistical properties of the radiation requires the determination of the hierarchy of distribution functions $p_n(\hat{E}_1, \dots, \hat{E}_n; x_1, \dots, x_n)$ introduced in Section IV.1. Due to the observed formal correspondence between the analytical signal $\hat{E}(\mathbf{r}, t)$ and the positive frequency part operator $E^+(\mathbf{r}, t)$ one would expect the quantum-mechanical case to be described in terms of a distribution function $p_n(\mathscr{E}_1, \mathscr{E}_2, \dots, \mathscr{E}_n)$ of the eigenvalues of $E^+(\mathbf{r}, t)$. Although, as we shall see, E^+ actually admits eigenvalues, it is a non-Hermitian operator with non-orthogonal eigenvectors, so that it is not possible to attach a precise meaning of probability to the $p_n(\mathscr{E}_1, \mathscr{E}_2, \dots, \mathscr{E}_n)$.

IV.6 The *P*-Representation

The formal structure of the quantum-mechanical correlation function $G^{(n)}$, as given by Eq. (4-75) through a well-ordered product of creation and annihilation operators, demonstrates the fundamental role of the eigenkets of the positive frequency part operator $E^{(+)}$.

The importance of these states in the frame of quantum

optics was first observed by Glauber (1963b) and consists mainly of the particularly simple expressions of the corresponding $G^{(n)}$'s. The introduction of these eigenkets has been made by Glauber via the so-called *coherent states*.

The *coherent states* $|\alpha\rangle$ are defined, for a single-mode case, in terms of the eigenstates $|n\rangle$ of the number operators [see Eq. (4-37)] as

$$|\alpha\rangle = \exp(-|\alpha|^2/2) \sum_{n=0}^{\infty} \frac{\alpha^n}{(n!)^{1/2}} |n\rangle \qquad (4\text{-}78)$$

and verify the eigenvalue equation

$$a|\alpha\rangle = \alpha|\alpha\rangle \qquad (4\text{-}79)$$

for any complex number α. They are not mutually orthogonal and form an overcomplete set, in the sense that every $|\alpha\rangle$ can be expanded as a linear combination of the others. The generalization to the multimode case is easily performed by introducing the coherent state $|\{\alpha_l\}\rangle$ defined as the product of the coherent states $|\alpha_l\rangle_l$ relative to each mode

$$|\{\alpha_l\}\rangle = \prod_l |\alpha_l\rangle_l \qquad (4\text{-}80)$$

It is easy to show that they are eigenkets of the positive frequency part of the electric field operator

$$E^{(+)}(\mathbf{r}, t)|\{\alpha_l\}\rangle = \mathscr{E}_{\{\alpha_l\}}(\mathbf{r}, t)|\{\alpha_l\}\rangle \qquad (4\text{-}81)$$

with complex eigenvalue $\mathscr{E}_{\{\alpha_l\}}(\mathbf{r}, t)$ given by

$$\mathscr{E}_{\{\alpha_l\}}(\mathbf{r}, t) = i \sum_l \left(\frac{\hbar\omega_l}{L^3}\right)^{1/2} \alpha_l \exp[i(\mathbf{k}_l \cdot \mathbf{r} - \omega_l t)] \qquad (4\text{-}82)$$

and, consequently, that

$$\langle\{\alpha_l\}| E^{(-)}(\mathbf{r}, t) = \mathscr{E}^*_{\{\alpha_l\}}(\mathbf{r}, t) \langle\{\alpha_l\}| \qquad (4\text{-}83)$$

According to Eqs. (4-81) and (4-83) the correlation function $G^{(n)}$ defined by Eq. (4-75) assumes for a coherent state $|\{\alpha_l\}\rangle$

the particularly simple form

$$G^{(n)}(\mathbf{r}_1 t_1, \ldots, \mathbf{r}_n t_n, \mathbf{r}_n t_n, \ldots, \mathbf{r}_1 t_1)$$

$$= \langle \{\alpha_l\} | E^{(-)}(\mathbf{r}_1, t_1) \cdots E^{(-)}(\mathbf{r}_n, t_n) E^{(+)}(\mathbf{r}_n, t_n)$$

$$\times \cdots E^{(+)}(\mathbf{r}_1, t_1) | \{\alpha_l\} \rangle$$

$$= \mathscr{E}^*_{\{\alpha_l\}}(\mathbf{r}_1, t_1) \cdots \mathscr{E}^*_{\{\alpha_l\}}(\mathbf{r}_n, t_n) \mathscr{E}_{\{\alpha_l\}}(\mathbf{r}_n, t_n) \cdots \mathscr{E}_{\{\alpha_l\}}(\mathbf{r}_1, t_1) \qquad (4\text{-}84)$$

Thus in this case, for which no statistical uncertainty is present in our knowledge of the state of the system, the quantum-mechanical expression for $G^{(n)}$ is equivalent to the classical expression given by Eq. (4-76) for a prescribed electromagnetic field with analytic signal $\mathscr{E}_{\{\alpha_l\}}(\mathbf{r}, t)$. This equivalence holds independently from the average number of photons[†] associated with the state $|\{\alpha_l\}\rangle$, the ultimate difference between the quantum-mechanical and classical cases lying in the fact that $|\{\alpha_l\}\rangle$ is an eigenstate of $E^{(+)}(\mathbf{r}, t)$ and not of the electric field operator $E(\mathbf{r}, t) = E^+(\mathbf{r}, t) + E^{(-)}(\mathbf{r}, t)$.

The preceding analogy can be extended to the situation in which one deals with a statistical mixture of coherent states. To this end, we briefly recall the possibility of describing the most general statistical quantum system by means of the so-called *density matrix* ρ.[‡] This is a Hermitian operator, time-independent in the Heisenberg picture, such that the quantum and statistical expectation value of any operator A is expressed as

$$\{\langle A \rangle\}_{\text{av}} = \text{Tr}(\rho A) \qquad (4\text{-}85)$$

where the *trace* Tr of an operator O is defined as

$$\text{Tr}(O) = \sum_{|\Psi\rangle} \langle \Psi | O | \Psi \rangle \qquad (4\text{-}86)$$

[†] This number can be shown immediately to be given by $\sum_l |\alpha_l|^2$ if one takes into account Eq. (4-79) and its Hermitian conjugate when evaluating $\langle \{\alpha_l\} | \sum_l a_l^+ a_l | \{\alpha_l\} \rangle$.

[‡] For a concise description of the properties of ρ in connection with problems of quantum optics see, for example, the work of Louisell (1964, Chapter VI).

the $|\Psi\rangle$'s forming a set of states verifying the completeness relation

$$\sum_{|\Psi\rangle} |\Psi\rangle \langle\Psi| = 1 \qquad (4\text{-}87)$$

It is immediately apparent that for a system in a *pure* state, $|\psi\rangle$ corresponding to no statistical indetermination, the density matrix ρ is given by the operator $|\psi\rangle\langle\psi|$ so that, for a coherent state, one has

$$\rho = |\{\alpha_l\}\rangle\langle\{\alpha_l\}| \qquad (4\text{-}88)$$

More in general, we can consider the situation in which ρ is expressed as a linear superposition of operators $|\{\alpha_l\}\rangle\langle\{\alpha_l\}|$ in the form

$$\rho = \int P(\{\alpha_l\})|\{\alpha_l\}\rangle\langle\{\alpha_l\}| \, d^2\{\alpha_l\} \qquad (4\text{-}89)$$

with $d^2\{\alpha_l\} \equiv \prod_l d(\mathrm{Re}\,\alpha_l)\, d(\mathrm{Im}\,\alpha_l)$. This representation for the density operator was introduced by Glauber (1963b) and Sudarshan (1963) and is known as *P-representation.*

In order to preserve the Hermitian and unitary character of the operator ρ, the *P*-representation must be real and satisfy the normalization condition

$$\int P(\{\alpha_l\})\, d^2\{\alpha_l\} = 1 \qquad (4\text{-}90)$$

while it may take on negative values. The possibility of expressing the density matrix in terms of the *P*-representation actually exists for a large class of physical situations and allows us to write the correlation function $G^{(n)}$ as

$$G^{(n)}(\mathbf{r}_1 t_1, \ldots, \mathbf{r}_n t_n, \mathbf{r}_n t_n, \ldots, \mathbf{r}_1 t_1)$$

$$= \mathrm{Tr}\,[\rho E^{(-)}(\mathbf{r}_1, t_1) \cdots E^{(-)}(\mathbf{r}_n, t_n)\, E^{(+)}(\mathbf{r}_n, t_n) \cdots E^{(+)}(\mathbf{r}_1, t_1)]$$

$$= \int P(\{\alpha_l\})|\mathscr{E}_{\{\alpha_l\}}(\mathbf{r}_1, t_1)|^2 \cdots |\mathscr{E}_{\{\alpha_l\}}(\mathbf{r}_n, t_n)|^2 \, d^2\{\alpha_l\} \qquad (4\text{-}91)$$

This expression is formally equivalent to what one would write in a classical situation for which Eq. (4-76) is valid, keeping in mind that one would deal with a nonnegative weight function having the meaning of probability density.

These considerations make clear the important role played by the P-representation, which, whenever it exists, allows us to perform in a formally identical way classical and quantum-mechanical calculations relevant to the evaluation of the hierarchy of the correlation functions $G^{(n)}$. In any event, the $P(\{\alpha_l\})$ cannot be interpreted, also when it assumes only positive values, as a probability density of finding the system in a given state $|\{\alpha_l\}\rangle$, since the coherent states are not mutually orthogonal. This interpretation becomes practically valid in the classical limit of large number of photons in which $P(\{\alpha_l\})$ tends to coincide with the corresponding probability density of finding a set of mode amplitudes $\{\alpha_l\}$, the structure of the coherent states being such that $\langle\{\alpha_l\}|\{\alpha_l'\}\rangle$ approaches zero in the classical limit for $\{\alpha_l\} \neq \{\alpha_l'\}$ (Glauber, 1963b, 1966).

IV.7 Photon-Counting Experiments

The statistical properties of electromagnetic radiation are actually tested by means of photon-counting devices, which contain, in practice, a very large number N of active atoms. The considerations of Section IV.5, concerning the n-atom photon detector, have to be generalized to deal with this case. More often, one looks for the probability $p(m, t)$ of counting m photons in a given time interval $0-t$ (where m is any integer smaller or equal to N), which is a basic possible photon-counting measurement.

A very elegant way of expressing $p(m, t)$ in terms of the statistical properties of the field hinges on the introduction of the *generating function* $Q(\lambda, t)$ (Glauber, 1965, 1967) defined through the expectation value

$$Q(\lambda, t) = \mathrm{Tr}\,[\rho(1 - \lambda)^{C(t)}] = \langle(1 - \lambda)^{C(t)}\rangle \qquad (4\text{-}92)$$

where $C(t)$ is the operator number of photons registered in the time interval $(0, t)$. Accordingly, the meaning of $p(m, t)$ allows us to write

$$Q(\lambda, t) = \sum_{m=0}^{\infty} (1 - \lambda)^m p(m, t) \qquad (4\text{-}93)$$

from which it follows that

$$p(m, t) = \frac{(-1)^m}{m!}\left[\frac{d^m}{d\lambda^m} Q(\lambda, t)\right]_{\lambda=1} \qquad (4\text{-}94)$$

Denoting by c_j the operator analogous to C pertaining to the jth atom, which obviously has eigenvalues 0 and 1, one has

$$(1 - \lambda)^C = (1 - \lambda)^{\sum c_j} = \prod_j (1 - \lambda)^{c_j} = \prod_j (1 - \lambda c_j) \qquad (4\text{-}95)$$

so that

$$Q(\lambda, t) = \left\langle \prod_j (1 - \lambda c_j) \right\rangle = 1 - \lambda \sum_j \langle c_j \rangle + \lambda^2 \sum_{j \neq l} \langle c_j c_l \rangle + \cdots \qquad (4\text{-}96)$$

where the sums have to be taken over all the possible combinations of atoms. The general term $\langle c_1 \cdots c_n \rangle$ on the right-hand side of Eq. (4-96) represents the probability that n photons are registered by a specified set of n atoms of· the counter, so that it coincides with $p^{(n)}(t)$ as given by Eq. (4-74), a quantity depending on the positions $\mathbf{r}_1, \mathbf{r}_2, \ldots, \mathbf{r}_n$ of the n atoms. This allows us to express the generic sum on the right-hand side of Eq. (4-96) as a sum over the positions of all the N atoms of the counter in the approximate form

$$\sum_{j_1 \neq j_2 \neq \cdots \neq j_n} \langle c_{j_1} c_{j_2} \cdots c_{j_n} \rangle \simeq \frac{1}{n!} \sum_{j_1=1}^{N} \cdots \sum_{j_n=1}^{N} p^{(n)}(t; \mathbf{r}_{j_1}, \mathbf{r}_{j_2}, \ldots, \mathbf{r}_{j_n}) \qquad (4\text{-}97)$$

where the $1/n!$ factor takes into account the fact that each set of n different atoms has to be considered once and only once. Each sum on the right-hand side of Eq. (4-97) can be expressed as an integral over the volume V_c of the counter, by introducing the microscopic atom density $\sigma(\mathbf{r})$:

$$\sum_{j_1=1}^{N} \to \int_{V_c} d\mathbf{r}\, \sigma(\mathbf{r}) \tag{4-98}$$

Thus Eqs. (4-74), (4-97), and (4-98) allow us to write Eq. (4-96) in the form

$$Q(\lambda, t) = \sum_{n=0}^{\infty} \frac{(-\lambda)^n}{n!} \beta^n \int_0^t dt_1 \cdots \int_0^t dt_n \int_{V_c} d\mathbf{r}_1 \cdots \int_{V_c} d\mathbf{r}_n$$

$$\times\, G^{(n)}(\mathbf{r}_1\, t_1, \ldots, \mathbf{r}_n\, t_n, \mathbf{r}_n\, t_n, \ldots, \mathbf{r}_1\, t_1) \tag{4-99}$$

having assumed a constant density σ and having defined $\beta = \sigma s$. While Eqs. (4-94) and (4-99) show that $p(m, t)$ contains global information on the $G^{(k)}$'s with $k \geqslant m$, direct information regarding the single-order k can be achieved by introducing the factorial moments M_k defined as

$$M_k = \langle C(C-1) \cdots (C-k+1) \rangle$$

$$= \sum_{m=0}^{\infty} m(m-1) \cdots (m-k+1)\, p(m, t) \tag{4-100}$$

In fact, Eqs. (4-93) and (4-100) immediately give

$$M_k = (-1)^k \left[\frac{d^k}{d\lambda^k} Q(\lambda, t) \right]_{\lambda=0} \tag{4-101}$$

so that

$$M_k = \beta^k \int_0^t dt_1 \cdots \int_0^t dt_k \int_{V_c} d\mathbf{r}_1 \cdots \int_{V_c} d\mathbf{r}_k$$

$$\times\, G^{(k)}(\mathbf{r}_1\, t_1, \ldots, \mathbf{r}_k\, t_k, \mathbf{r}_k\, t_k, \ldots, \mathbf{r}_1\, t_1) \tag{4-102}$$

having taken into account Eq. (4-99).

Equation (4-94) can be cast in a significant form by resorting to the *P*-representation. In fact, by inserting the expression for $G^{(n)}$, as given by Eq. (4-91), into Eq. (4-99), one easily obtains

$$Q(\lambda, t) = \int P(\{\alpha_k\}) \exp[-\lambda\Omega(\{\alpha_k\})] \, d^2\{\alpha_k\} \qquad (4\text{-}103)$$

where

$$\Omega(\{\alpha_k\}) = \beta \int_0^t dt' \int_{V_c} d\mathbf{r}' \, \mathscr{E}^*_{\{\alpha_k\}}(\mathbf{r}', t') \mathscr{E}_{\{\alpha_k\}}(\mathbf{r}', t') \qquad (4\text{-}104)$$

so that

$$p(m, t) = \int P(\{\alpha_k\}) \frac{\Omega^m(\{\alpha_k\})}{m!} \exp[-\Omega(\{\alpha_k\})] \, d^2\{\alpha_k\} \qquad (4\text{-}105)$$

Equation (4-105) represents the fully quantum-mechanical expression for photon-counting distribution over a finite time interval (Glauber, 1965; Kelley and Kleiner, 1964). Its classical limit assumes a simple analytical form for the case in which the quantity $\mathscr{E}^*_{\{\alpha_k\}}(\mathbf{r}, t) \mathscr{E}_{\{\alpha_k\}}(\mathbf{r}, t)$ does not significantly vary on the volume of the counter. It was first deduced by Mandel (1958) as

$$p(m, t) = \frac{1}{m!} \langle [\eta U(t)]^m e^{-\eta U(t)} \rangle \qquad (4\text{-}106)$$

where η is the *quantum efficiency* (Mandel and Wolf, 1965) of the counter and $U(t)$ is expressed in terms of the instantaneous intensity $i(t)$ as

$$U(t) = \int_0^t I(t') \, dt' \qquad (4\text{-}106)'$$

In particular, whenever $I(t')$ is nearly constant in the time interval $(0, t)$, Eq. (4-106) can be written in terms of the probability distribution $P_0(I)$ of the intensity as

$$p(m, t) = \int_0^{+\infty} P_0(I) \frac{(\eta I t)^m}{m!} e^{-\eta I t} \, dI \qquad (4\text{-}107)$$

We wish to remark that the possibilities of examining the statistical properties of an electromagnetic radiation field are not necessarily exhausted by the measurement of the photon-counting distribution $p(m, t)$. A number of alternative measurements can be performed, as we shall see in the following when dealing with specific cases concerning scattered radiation.

IV.8 Photon-Counting Distribution for Some Typical Fields

Most of the light sources occurring in nature present a common feature consisting in the possibility of representing their macroscopic radiated field as the sum of many similar, statistically independent, microscopic contributions. It is possible to characterize this kind of field by means of a Gaussian form of the corresponding P-representation as (Glauber, 1963b)

$$P(\{\alpha_k\}) = \prod_k \frac{1}{\langle n_k \rangle} \exp\left(-\frac{|\alpha_k|^2}{\langle n_k \rangle}\right) \qquad (4\text{-}108)$$

where $\langle n_k \rangle$ is the mean number of photons pertaining to the k-mode. This distribution seems the most appropriate to describe the electromagnetic field radiated by usual light sources, such as gas discharges, lamps, and so on, a type of radiator usually termed *chaotic*. A relevant example of chaotic electromagnetic radiation is furnished by *black-body radiation*, which possesses a P-representation of the form given in Eq. (4-108), with

$$\langle n_k \rangle = \frac{1}{\{\exp[\hbar\omega_k/(K_B T)] - 1\}} \qquad (4\text{-}109)$$

as can be directly shown starting from the well-known expression of the density matrix for a quantum-mechanical system at thermal equilibrium (Mandel and Wolf, 1965).

The correlation functions $G^{(n)}(\mathbf{r}_1 t_1, \ldots, \mathbf{r}_n t_n, \mathbf{r}_n t_n, \ldots, \mathbf{r}_1 t_1)$ associated with a chaotic field obey the relation

$$G^{(n)}(\mathbf{r}_1\,t_1, \ldots, \mathbf{r}_n\,t_n, \mathbf{r}_n\,t_n, \ldots, \mathbf{r}_1\,t_1) = \sum_{(p)} \prod_{i,j=1}^{n} G^{(1)}(\mathbf{r}_i\,t_i, \mathbf{r}_j\,t_j) \quad (4\text{-}110)$$

where the sum is taken over the $n!$ permutations of the index $j (j = 1, 2, \ldots, n)$ as can be seen by inserting Eq. (4-108) into Eq. (4-91) (Glauber, 1963*b*).

The photon-counting distribution relative to chaotic fields can, in principle, be evaluated by means of Eqs. (4-108) and (4-105) (Glauber, 1965). It assumes a particularly simple form whenever the counting time t is much less than the inverse frequency bandwidth of the radiation field, that is,

$$p(m, t) = \frac{\langle m \rangle^m}{(1 + \langle m \rangle)^{m+1}} \quad (4\text{-}111)$$

This formula is called the *Bose–Einstein distribution* since, as a particular relevant case, it describes the probability distribution of the occupation number for a boson gas at thermal equilibrium, whenever a single mode is considered (see Fig. 4.2).

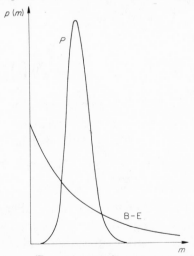

Fig. 4.2 *Typical photon-counting distribution: Bose–Einstein (B–E) and Poisson* (*P*).

The factorial moments M_k pertaining to the distribution given in Eq. (4-111) are

$$M_k = k! \langle m \rangle^k \qquad (4\text{-}112)$$

Another relevant situation is the one in which the correlation functions $G^{(n)}$ factor according to

$$G^{(n)}(\mathbf{r}, t_1, \ldots, \mathbf{r}_n t_n, \mathbf{r}_n t_n, \ldots, \mathbf{r}_1 t_1) = \prod_{j=1}^{n} G^{(1)}(\mathbf{r}_j t_j, \mathbf{r}_j t_j) \qquad (4\text{-}113)$$

This property holds true for fields emitted by sources that can be considered, in a sense, opposite to the chaotic ones. In fact, remarkable examples of this type are given by the radiation generated by a predetermined charge–current distribution (like, e.g., in an antenna; see Glauber, 1965) and by an ideal single-mode laser. In both cases, the microsources responsible for the field act in a strictly correlated way, thus presenting a completely different behavior from the chaotic ones.

The P-representation is of the form

$$P(\{\alpha_k\}) = \prod_{k} \delta^{(2)}(\alpha_k - \alpha_{k_0}) \qquad (4\text{-}114)$$

and

$$P(\{\alpha_k\}) = \frac{1}{2\pi |\alpha_{k'}|} \delta(|\alpha_{k'}| - |\alpha_{k_0}|) \prod_{k \neq k'} \delta^{(2)}(\alpha_k) \qquad (4\text{-}115)$$

respectively, for the field emitted by an antenna and a laser operating in the single mode k'_0, where the two-dimensional δ-function of complex argument $\delta^{(2)}(z)$ is defined as

$$\delta^{(2)}(z) = \delta(\mathrm{Re}\, z)\, \delta(\mathrm{Im}\, z) \qquad (4\text{-}116)$$

Equations (4-114) and (4-115) are both consistent with Eq. (4-113); Eq. (4-114) characterizes a coherent state of the field with eigenvalue $\mathscr{E}_{\{\alpha_{k_0}\}}(\mathbf{r}, t)$ [see Eq. (4-81)], while Eq. (4-115) describes a completely random mixture of coherent states having the same average number of photons $|\alpha_{k_0}|^2$ and all possible values of the phase in the interval $(0, 2\pi)$.

The photon-counting distribution associated with Eq. (4-113) is a *Poisson distribution* (Fig. 4.2)

$$p(m, t) = \frac{\langle m \rangle^m}{m!} e^{-\langle m \rangle} \tag{4-117}$$

and the corresponding factorial moments obey the relation

$$M_k = \langle m \rangle^k \tag{4-118}$$

V

Statistical Properties of Light Scattered by Fluids and Plasmas

V.1 The Role of Higher-Order Correlation Functions

In the previous chapter, the characterization of an electromagnetic radiation by means of its higher-order statistical properties was examined, and we showed how photon counting provides a convenient method for measuring these properties. In particular, the microscopic structure of a source has a direct influence on the statistics of the emitted radiation, which in turn can furnish the possibility of tracing back the corresponding emission process.

In this respect, we have already noted the difference between the structure of the hierarchies of correlation functions associated with lasers and conventional light sources, which corresponds to different behaviors for the photon-counting distributions. These are two typical situations for which all physical information is contained in the first-order correlation function $G^{(1)}$, since the higher-order correlation functions $G^{(n)}$ can be expressed in terms of $G^{(1)}$ [see Eqs. (4-110) and (4-113)]. It is clear that two sources with the same factorization properties describing a similar general statistical behavior can still be distinguished because of the different values of $G^{(1)}$. As an example, although both black-body radiation and *Bremsstrahlung* by a stable plasma are completely chaotic,[†] they differ deeply in their spectral properties or, equivalently, in the temporal behavior of $G^{(1)}$. Conversely, two fields possessing the same spectral distribution can actually differ for their higher-order statistical properties, this being the case, for example, for a single-mode laser field and a filtered chaotic radiation with the same frequency spectrum.

According to the preceding considerations, it is clear that a detailed knowledge of all-order correlations $G^{(n)}$ can yield information on the dynamics of a macroscopic light source. In this respect, much attention has been devoted to the limiting cases of chaotic and coherent sources. As we have already seen in Chapter IV, all sources consisting of many similar independent radiators belong to the first category. Another large class of this type is given by the material systems possessing *linear constitutive relations* (Korenman, 1967), although it has not been established if linearity and independence of microsources are complementary descriptions of the same physical situation. The laser has been extensively studied by many authors, from the point of view of its higher-order coherence properties, and the reader is referred to the book by Haken (1970) for a comprehensive theoretical treatment. Essentially, the *superradiating* systems belong to the same category of sources, where a cooperative spontaneous emission from an ensemble of N identical two-level atoms takes place. Under suitable conditions for the initial state of the system of atoms, the statistics of the emitted radiation

---

[†] A direct demonstration of the chaotic nature of *Bremsstrahlung* has been given by Crosignani *et al* (1968b).

tend to become coherent (see, for example, Bonifacio and Preparata, 1970, and Agarwal 1970).

The great majority of material systems of interest does not, however, under normal conditions, emit radiation. Thus they must, in a sense, be made to behave as light sources in order to extract information from the analysis of the collected radiation. This is precisely the role of a scattering experiment.

This type of experiment has undergone a rapid development due to the possibility of employing highly stabilized laser sources, which are able to produce electromagnetic fields practically free of statistical uncertainty. As we shall see, this allows us to establish very simple formal relations between the statistical properties of scattered light and those describing the scattering medium. They will be shown to connect higher-order correlation functions of the medium to those of scattered light, in such a way as to generalize (in a sense that will be made precise later) the well-known relation between scattering cross section and lowest-order correlation function of the fluctuating system.

V.2 Correlation Functions of Light Scattered by a Simple Fluid

The most relevant quantity usually measured in a conventional scattering experiment is the *differential scattering cross section per unit frequency* $d\sigma/d\Omega\, d\omega$, defined as the ratio between the electromagnetic power emitted per unit solid angle and frequency in a given direction, and the incident intensity. Accordingly, one has

$$\frac{d\sigma}{d\omega} = \frac{I(\omega, \eta)\, r^2\, d\Omega}{(c/8\pi)\, E_0^2} \tag{5-1}$$

where $I(\omega, \eta)$ is the power spectrum of the scattered light in the direction η at a distance r from the scattering volume V_{sc} such that $r \gg V_{sc}^{1/3}$, and $d\Omega$ is the element of solid angle around the direction η, the incident field being represented by the analytic signal (see Fig. 5.1)

$$\hat{\mathbf{E}}(\mathbf{r}, t) = \mathbf{E}_0 \exp[i(\mathbf{k}_0 \cdot \mathbf{r} - \omega_0 t)] \qquad (5\text{-}2)$$

By inserting the single-scattering expression of $I(\omega, \boldsymbol{\eta})$ into Eq. (5-1), for example the microscopic expression deduced in Chapter II, one immediately obtains the relation between cross section and Fourier space–time transform of the two-particle correlation function $\mathscr{G}_2(\mathbf{r}, t)$ [see Eqs. (2-35) and (2-59)], in the form

$$\frac{d\sigma}{d\Omega\, d\omega} = k_0^4 \alpha^2(\omega_0)\, N^2 V_{sc} \left| \boldsymbol{\eta} \times \left(\boldsymbol{\eta} \times \frac{\mathbf{E}_0}{E_0} \right) \right|^2 \tilde{\mathscr{G}}_2(-\mathbf{k}_1, \omega - \omega_0)$$

$$(5\text{-}3)$$

This is the basic equation relating the measured cross section to the two-particle correlation function, which is the lowest-order statistical description of a homogeneous stationary fluctuating medium. In other words, a simple relation exists between the lowest-order statistical properties of the scattering system and those of the scattered radiation. Thus it is natural to look for the

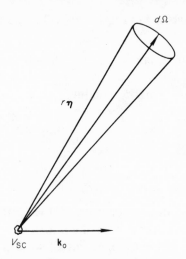

Fig. 5.1   *Scattering geometry.*

possibility of obtaining wider information by using the general description of a stochastic electromagnetic field introduced in Chapter IV.

To this end, we must evaluate the nth order correlation function $G^{(n)}$ relative to the scattered electric field. This field will be treated in the form of a classical random variable [a quantum-mechanical treatment of light scattered by a fluctuating medium has been considered by Hellwarth (1970)], since we shall consider incident and scattered electric fields large enough to be excited to classical intensities. We shall employ Eq. (4-76) using mainly the analytic signal $\hat{\mathbf{E}}_{sc}(\mathbf{r}, t)$ corresponding to the singly scattered electric field given by Eq. (1-89), although the following developments can be easily modified by employing the macroscopic expression of the scattered field (which can include multiple-scattering contributions; see Chapter III).

It is convenient to express $\hat{\mathbf{E}}_{sc}(\mathbf{r}, t)$ in terms of the Fourier space transform $\tilde{n}_1(\mathbf{k}, t)$ of the microscopic density [see Eq. (2-27)], that is,

$$\hat{\mathbf{E}}_{sc}(\mathbf{r}, t) = -\frac{k_0^2 \alpha(\omega_0)}{r} \exp\left[-i\omega_0\left(t - \frac{r}{c}\right)\right] \boldsymbol{\eta} \times (\boldsymbol{\eta} \times \mathbf{E}_0)$$

$$\times \tilde{n}_1\left(-\mathbf{k}_1, t - \frac{r}{c}\right) \tag{5-4}$$

where

$$\tilde{n}_1(\mathbf{k}, t) = \int_{V_{sc}} n_1(\mathbf{r}', t) e^{-i\mathbf{k}\cdot\mathbf{r}'} d\mathbf{r}' \tag{5-5}$$

If we consider photon-counting measurements in which a single counter is employed, such that the intensity of the scattered field does not appreciably vary on its active surface, we need only evaluate correlation functions of the kind

$$G^{(n)}(\mathbf{r}t_1, \ldots, \mathbf{r}t_n, \mathbf{r}t_n, \ldots, \mathbf{r}t_1)$$

in which $\mathbf{r}$ labels the mean position of the counter. By inserting

Eq. (5-4) into Eq. (4-76), we obtain

$$G^{(n)}(\mathbf{r}t_1, \ldots, \mathbf{r}t_n, \mathbf{r}t_n, \ldots, \mathbf{r}t_1)$$

$$= \gamma^n \langle \tilde{n}_1(\mathbf{k}_1, t_1) \tilde{n}_1(-\mathbf{k}_1, t_1) \cdots \tilde{n}_1(\mathbf{k}_1, t_n) \tilde{n}_1(-\mathbf{k}_1, t_n) \rangle$$

$$(5\text{-}6)$$

where

$$\gamma = \frac{k_0^4 \alpha^2(\omega_0)}{r^2} |\boldsymbol{\eta} \times (\boldsymbol{\eta} \times \mathbf{E}_0)|^2 \qquad (5\text{-}7)$$

having taken into account the relation $\tilde{n}_1^*(\mathbf{k}, t) = \tilde{n}_1(-\mathbf{k}, t)$, which is a consequence of the reality of $n_1(\mathbf{r}, t)$. Besides, all times $t_i - (r/c)$ have been translated by an amount r/c, this being justified by the stationary hypothesis on the scattering system. The further hypothesis of homogeneity implies

$$n_1(\mathbf{r}, t) = \langle n_1 \rangle + \Delta n_1(\mathbf{r}, t) \qquad (5\text{-}8)$$

where the ensemble average $\langle n_1 \rangle$ does not depend on space, so that the set of Eqs. (5-6) can be rewritten as

$$G^{(n)}(\mathbf{r}t_1, \ldots, \mathbf{r}t_n, \mathbf{r}t_n, \ldots, \mathbf{r}t_1)$$

$$= \gamma^n \int_{V_{sc}} d\mathbf{r}_1 \int_{V_{sc}} d\mathbf{r}_1' \cdots \int_{V_{sc}} d\mathbf{r}_n \int_{V_{sc}} d\mathbf{r}_n'$$

$$\times \exp[i\mathbf{k}_1 \cdot (\mathbf{r}_1 - \mathbf{r}_1' + \cdots + \mathbf{r}_n - \mathbf{r}_n')]$$

$$\times \langle \Delta n_1(\mathbf{r}_1, t_1) \Delta n_1(\mathbf{r}_1' \, t_1) \cdots \Delta n_1(\mathbf{r}_n, t_n) \Delta n_1(\mathbf{r}_n' \, t_n) \rangle$$

$$(5\text{-}9)$$

or, recalling Eq. (2-36),

$$G^{(n)}(\mathbf{r}t_1, \ldots, \mathbf{r}t_n, \mathbf{r}t_n, \ldots, \mathbf{r}t_1)$$

$$= (\gamma N^2)^n \int_{V_{sc}} d\mathbf{r}_1 \int_{V_{sc}} d\mathbf{r}_1' \cdots \int_{V_{sc}} d\mathbf{r}_n \int_{V_{sc}} d\mathbf{r}_n'$$

$$\times \exp[i\mathbf{k}_1 \cdot (\mathbf{r}_1 - \mathbf{r}_1' + \cdots + \mathbf{r}_n - \mathbf{r}_n')]$$

$$\times \mathscr{G}_{2n}(\mathbf{r}_1, \mathbf{r}_1', \ldots, \mathbf{r}_n, \mathbf{r}_n', t_1, t_1, \ldots, t_n, t_n) \qquad (5\text{-}10)$$

The set of Eqs. (5-10), or equivalently of Eq. (5-6) or (5-9), provides a straightforward relation between the correlation functions of the scattered field and those of the scattering medium. More precisely, the nth order correlation function of the field scattered at a given point $\mathbf{r}$ only depends on the $2n$th order correlation function of the scattering system through a particular Fourier transform in space. Thus a measurement of

$$G^{(n)}(\mathbf{r}t_1, \ldots, \mathbf{r}t_n, \mathbf{r}t_n, \ldots, \mathbf{r}t_1)$$

yields, for every n, the possibility of gaining information on the ensemble average of products of more than two density fluctuations, in contrast with an ordinary cross-sectional measurement. The overall structure of the hierarchy of the $\mathscr{G}_{2n}$'s is in turn directly related to the fundamental dynamic properties of the system.

Obviously, whenever the scattering process is due to non-fluctuating inhomogeneities of the medium rather than to statistical fluctuations of density, no uncertainty is added to that possessed by the incident electromagnetic radiation, so that the structure of the hierarchy of the $G^{(n)}$'s does not undergo changes due to the scattering process. Besides, there are a number of cases in which, if the incident light is coherent, i.e., its $G^{(n)}$'s factor according to Eq. (4-113), the correlation functions of the scattered field obey Eq. (4-110), which is characteristic of a chaotic radiation. While in these two limiting situations all the $G^{(n)}$'s can be expressed in terms of the $G^{(1)}$, there are relevant cases in which this is no longer true and the measurement of some specific $G^{(n)}$ with $n > 1$ yields direct information on the medium not contained in the $G^{(1)}$.

The suggestion of measuring second-order properties of the scattered field in order to obtain information not available with an ordinary first-order cross-sectional experiment was first made by Goldberger *et al.* (1963). Their analysis was mainly concerned with the scattering of beams of ordinary particles, while the specific case of a photon beam has been successively examined by Fetter (1965). These treatments are devoted to the investigation of particular features of the target, knowledge of which can be gained, for example, by measuring the phase of the scattering amplitude, besides its modulus, which is directly related to the angular cross section. These authors do not deal, however, with the determination of the statistical properties of the scattering system, which are assumed to be known a priori.

The general formal relation between higher-order correlation functions of the scattering system and those of the scattered field has been examined, in connection with the experimental potentialities offered by photon-counting technique, by Crosignani and Di Porto (1967), Bertolotti *et al.* (1967), and Shen (1967, 1969).

Before developing in some detail the arguments set forth in this section it is worth describing some kinds of higher-order measurements that can be performed on the electromagnetic field, different from those examined in the previous chapter.

V.3   Correlation Functions of the Scattered Field:
   Other Kinds of Measurements

Besides the photon-counting experiments performed with a single broadband counter, as described in Section IV.6, there are other measurements in which the observable quantities are connected with correlation functions of the electromagnetic field different from those defined in Eq. (4-76), which are a particular case of the more general class of correlation functions defined as

$$G^{(n,m)}(\mathbf{r}_1 t_1, \ldots, \mathbf{r}_n t_n, \mathbf{r}_{n+1} t_{n+1}, \ldots, \mathbf{r}_{n+m} t_{n+m})$$
$$= \langle \hat{E}^*(\mathbf{r}_1, t_1) \cdots \hat{E}^*(\mathbf{r}_n, t_n) \hat{E}(\mathbf{r}_{n+1}, t_{n+1}) \cdots \hat{E}(\mathbf{r}_{n+m}, t_{n+m}) \rangle$$
$$(5\text{-}11)$$

It is evident that a measurement of the $G^{(n,m)}$'s would bring more complete information on the scattering medium. Although it is possible, at least in principle, to measure $G^{(n,m)}$ with $n \neq m$

by taking advantage of suitable nonlinear processes (Glauber, 1969, 1970; Peřina, 1971, Chapter XII), the level of power of scattered radiation does not allow us to consider them as very practical for our purposes, so that it is more realistic to consider the measurement of

$$G^{(n,n)}(\mathbf{r}t_1, \ldots, \mathbf{r}t_n, \mathbf{r}t_{n+1}, \ldots, \mathbf{r}t_{2n})$$

$$= G^{(n)}(\mathbf{r}t_1, \ldots, \mathbf{r}t_n, \mathbf{r}t_{n+1}, \ldots, \mathbf{r}t_{2n}) \qquad (5\text{-}12)$$

which is a generalization of Eq. (4-76) due to the arbitrariness of the temporal arguments.

We have already seen that a spectral measurement is directly related to the correlation function

$$G^{(1)}(\mathbf{r}t, \mathbf{r}t') = \langle \hat{E}^*(\mathbf{r}, t)\,\hat{E}(\mathbf{r}, t')\rangle \qquad (5\text{-}13)$$

by means of the Wiener–Khintchine theorem [see Eq. (2-48)]. The direct generalization of this quantity to the successive order is

$$G^{(2)}(\mathbf{r}t, \mathbf{r}t', \mathbf{r}t'', \mathbf{r}t''') = \langle \hat{E}^*(\mathbf{r}, t)\,\hat{E}^*(\mathbf{r}, t')\,\hat{E}(\mathbf{r}, t'')\,\hat{E}(\mathbf{r}, t''')\rangle$$

$$(5\text{-}14)$$

for which a method of measurement has been devised by Cantrell (1968). (A different method based on the process of second harmonic generation has been proposed by Beran *et al.* (1967), but it requires high levels of electromagnetic power.) The main idea consists of measuring the intensity correlation of the field after a suitable filtering operation. More precisely (see Fig. 5.2), the electromagnetic beam is split into two beams that are in turn made to pass through a filtering device (Fabry–Perot interferometer, grating spectrometer) and then collected by two detectors whose outputs are sent to an electronic device (correlator) capable of performing ensemble averages of their product, thus recording a quantity proportional to

$$\langle I_{F_1}(t)\,I_{F_2}(0)\rangle = \left(\frac{c}{8\pi}\right)^2 \langle\,|\hat{E}_{F_1}(t)|^2|\hat{E}_{F_2}(0)|^2\rangle \qquad (5\text{-}15)$$

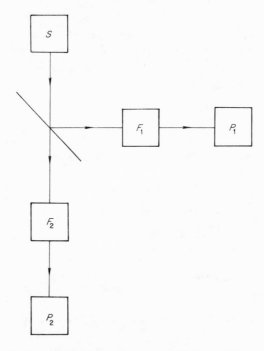

Fig. 5.2 *Measurement of the temporal correlation of light (after Cantrell, 1968).*

The analytic signals $\hat{E}_{F_1}(t)$ and $\hat{E}_{F_2}(t)$, representing the two filtered fields, can be expressed in terms of the response functions $F_1(\omega) = F_1(\omega - \Omega_1)$ and $F_2(\omega) = F_2(\omega - \Omega_2)$ of the two filters as

$$\hat{E}_{F_j}(t) = \int_{-\infty}^{+\infty} \Psi_j(\omega) F_j(\omega - \Omega_j) e^{-i\omega t}\, d\omega, \qquad j = 1, 2$$

$$(5\text{-}16)$$

the analytic signals of the two fields before filtering being represented by

$$\hat{E}_j(t) = \int_{-\infty}^{+\infty} \Psi_j(\omega) e^{-i\omega t}\, d\omega, \qquad j = 1, 2 \qquad (5\text{-}17)$$

with $\Psi_j(\omega) = 0$ for $\omega < 0$.

The relevant quantity that has to be considered is the Fourier time transform $\tilde{\mathscr{I}}(\Omega_1, \Omega_2, \omega)$ of the normalized intensity correlation

$$\mathscr{I}(\Omega_1, \Omega_2, t) = \frac{\langle |\hat{E}_{F_1}(t)|^2 |\hat{E}_{F_2}(0)|^2 \rangle}{\langle |\hat{E}_{F_1}(t)|^2 \rangle \langle |E_{F_2}(t)|^2 \rangle} \qquad (5\text{-}18)$$

After some algebraic manipulations, one obtains for a stationary situation

$$\tilde{\mathscr{I}}(\Omega_1, \Omega_2, \omega)$$

$$= \frac{1}{\langle |\hat{E}_{F_1}(t)|^2 \rangle \langle |\hat{E}_{F_2}(t)|^2 \rangle} \frac{1}{2\pi} \int_{-\infty}^{+\infty} dt_1 \int_{-\infty}^{+\infty} dt_2 \int_{-\infty}^{+\infty} dt_3$$

$$\times K_1(\omega, t_1) K_2(-\omega, t_2 - t_3)$$

$$\times \exp\{-i[(\Omega_1 + \omega) t_1 + (\Omega_2 - \omega) t_2 - \Omega_2 t_3]\}$$

$$\times \langle \hat{E}_1^*(t_1) \hat{E}_2^*(t_2) \hat{E}_2(t_3) \hat{E}_1(0) \rangle \qquad (5\text{-}19)$$

having defined the convolution integral

$$K_j(\omega, t) = \int_{-\infty}^{+\infty} \tilde{F}_j^*(-u) \tilde{F}_j(t-u) e^{-i\omega u} \, du \qquad (5\text{-}20)$$

of the Fourier transform $\tilde{F}_j(u)$ of $F_j(\omega)$. Equation (5-19) can be cast in a more convenient form by means of the change of variables $\tau_1 = t_1$, $\tau_2 = t_2 - t_3$, $\tau_3 = t_1 - t_2$. This allows us to perform a Fourier inversion, so that we can write the correlation function of the field as

$$G^{(2)}(t_1, t_2, t_3, 0)$$

$$= \langle I_{F_1}(t) \rangle \langle I_{F_2}(t) \rangle D \int_{-\infty}^{+\infty} d\Omega_1 \int_{-\infty}^{+\infty} d\Omega_2 \int_{-\infty}^{+\infty} d\omega$$

$$\times \frac{\exp\{i[\Omega_1 t_1 + \Omega_2 (t_2 - t_3) + \omega(t_1 - t_2)]\}}{K_1(\omega, t_1) K_2(-\omega, t_2 - t_3)}$$

$$\times \tilde{\mathscr{I}}(\Omega_1, \Omega_2, \omega) \qquad (5\text{-}21)$$

where D is a suitable constant depending on the experimental arrangement. If $\tilde{\mathscr{I}}(\Omega_1,\Omega_2,\omega)$ is measured as a function of the relevant values of the variable Ω_1,Ω_2,ω, Eq. (5-21) allows us in principle to determine the second-order correlation function $G^{(2)}(t_1,t_2,t_3,0)$ evaluated in the same space and at different times.

In practice the method is feasible if the bandwidth of the filters $\Delta\omega^{(F)}$ satisfies the relation $\Delta\omega^{(F)} < \Delta\omega^{(S)}$, where $\Delta\omega^{(S)}$ represents the bandwidth of the field, which amounts to saying that the product $K_1(\omega,t_1)K_2(-\omega,t_2-t_3)$ varies, as a function of t_1,t_2,t_3, less rapidly than the correlation function $G^{(2)}(t_1,t_2,t_3,0)$. In fact, this assures us that $K_1 K_2$ does not vanish for times in which $G^{(2)}(t_1,t_2,t_3,0)$ is still different from zero.

The particular class of $G^{(n)}$ given by Eq. (4-75), that is, those coinciding in the classical limit with the ensemble average of the product of n intensities evaluated at equal positions and different times [see Eq. (4-76)], can be measured by means of methods different from photon counting in a given time interval, as introduced in the preceding chapter. For example, the probability that if a photon has been recorded by a detector at time t a further photon is recorded at time $t+\tau$ within $d\tau$ is known as *conditional probability*, and is given by (see, for example, Mandel, 1967)

$$p_c(t,t+\tau)\,d\tau = \beta\frac{\langle I(\mathbf{r},t)\,I(\mathbf{r},t+\tau)\rangle}{\langle I(\mathbf{r},t)\rangle}\,d\tau \qquad (5\text{-}22)$$

where $\mathbf{r}$ is the (mean) position of the counter and β its photo-efficiency. One can consider as well the *intensity-fluctuation spectrum* defined, in a stationary case, as (see, for example, Rice, 1954)

$$P(\omega) = \frac{1}{2\pi}\int_{-\infty}^{+\infty}\langle I(\mathbf{r},t)\,I(\mathbf{r},t+\tau)\rangle\,e^{i\omega\tau} \qquad (5\text{-}23)$$

for all values of ω, which represents the natural extension of the ordinary concept of power spectrum of the electromagnetic field to the successive order.

One could consider at this point the symmetric problem of measuring the correlation function $G^{(n)}$ at different positions. Although a direct analog of the method previously outlined in the time domain has not been considered in order to measure nth order correlation functions with all different $\mathbf{r}_i$ $(i = 1, 2, ..., n)$, one can determine correlation functions of the kind given by Eq. (4-75) by placing n counters in n different positions and recording the corresponding coincidences. Thus, for example, Eq. (5-22) generalizes to the case in which the photon at time t is detected by a counter placed in $\mathbf{r}_1$, while the other photon is detected by a counter placed in $\mathbf{r}_2$, to yield

$$p_c(\mathbf{r}_1, t, \mathbf{r}_2, t+\tau)\, d\tau \;=\; \beta_2 \, \frac{\langle I(\mathbf{r}_1, t)\, I(\mathbf{r}_2, t+\tau)\rangle}{\langle I(\mathbf{r}_1, t)\rangle}\, d\tau \quad (5\text{-}24)$$

where β_2 represents the efficiency of the second counter.

Most attention has been devoted to photon counting in the time domain, that is, to the investigation of the temporal behavior of the $G^{(n)}$'s. No corresponding systematic experimental analysis has been carried out in the space domain, in order to determine the spatial behavior of the $G^{(n)}$'s, by using two or more photon counters and varying their mutual distances. This is due to the fact that in most practical situations (and particularly in scattering experiments) the available experimental technique is more likely adapted to temporal rather than spatial variations, at least on the scale in which they take place for the $G^{(n)}$'s.

V.4 Nature of the Information Contained in the Scattered Field

Higher-order measurements performed on the scattered field can yield information on the medium not available with a conventional first-order experiment, only if the $G^{(n)}$'s with $n > 1$ are not a priori deducible from $G^{(1)}$. In this respect, the less interesting situation is the one in which the density fluctuations of the scattering system are known to be distributed according to *Gaussian statistics*.

We remember that the distribution function of a set of real stochastic variables $Y_1, Y_2, ..., Y_n$ is *normal* or *Gaussian* if it is the Fourier transform of the characteristic function (Chandrasekhar, 1951; Rice, 1954):

$$Q(\lambda_1, \lambda_2, ..., \lambda_n) = \exp(-\tfrac{1}{2} \sum_{i=1}^{n} \sum_{l=1}^{n} \langle Y_i Y_l \rangle \lambda_i \lambda_l) \quad (5\text{-}25)$$

Equation (5-25) contains the Gaussian factorization property, according to which any average of the kind $\langle Y_1^{m_1} Y_2^{m_2} \cdots Y_n^{m_n} \rangle$ vanishes if $\sum_{i=1}^{n} m_i$ is odd, while it is equal to the sum of the $(\sum_{i=1}^{n} m_i - 1)!!$ products of the kind $\langle Y_i Y_l \rangle \cdots \langle Y_h Y_k \rangle$, which can be constructed by separating into couples the $\sum_{i=1}^{n} m_i$ factors appearing in $\langle Y_1^{m_1} Y_2^{m_2} \cdots Y_n^{m_n} \rangle$, whenever $\sum_{i=1}^{n} m_i$ is even. As an example,

$$\langle Y^{2n} \rangle = (2n-1)!! \langle Y^2 \rangle^n \quad (5\text{-}26)$$

and

$$\langle Y_1 Y_2 Y_3 Y_4 \rangle$$
$$= \langle Y_1 Y_2 \rangle \langle Y_3 Y_4 \rangle + \langle Y_1 Y_3 \rangle \langle Y_2 Y_4 \rangle + \langle Y_1 Y_4 \rangle \langle Y_2 Y_3 \rangle$$
$$(5\text{-}27)$$

If the set of any $2n$ density fluctuations $\Delta n_1(\mathbf{r}_i, t_i)$ obeys this kind of statistics, we have

$$\langle \Delta n_1(\mathbf{r}_1, t_1) \Delta n_1(\mathbf{r}_2, t_2) \cdots \Delta n_1(\mathbf{r}_{2n}, t_{2n}) \rangle$$

$$= \sum_{\substack{l_1, l_2, ..., l_{2n} \\ \text{all different}}} \langle \Delta n_1(\mathbf{r}_{l_1}, t_{l_1}) \Delta n_1(\mathbf{r}_{l_2}, t_{l_2}) \rangle$$
$$\cdots \langle \Delta n_1(\mathbf{r}_{l_{2n-1}}, t_{l_{2n-1}}) \Delta n_1(\mathbf{r}_{l_{2n}}, t_{l_{2n}}) \rangle, \quad 1 \leqslant l_i \leqslant 2n$$
$$(5\text{-}28)$$

where the sum is extended to the $(2n-1)!!$ possible products obtained by separating into couples the $\Delta n_1(\mathbf{r}_i, t_i)$ appearing on the left-hand side of Eq. (5-28).

If we now introduce Eq. (5-28) into Eq. (5-9), we obtain the following expression for the correlation function $G^{(n)}$ of the scattered field:

$$G^{(n)}(\mathbf{r}t_1, \ldots, \mathbf{r}t_n, \mathbf{r}t_n, \ldots, \mathbf{r}t_1)$$

$$= \sum_{(p)} \prod_{\substack{l=1 \\ i \text{ all different}}}^{n} \gamma \int_{V_{sc}} d\mathbf{r}' \int_{V_{sc}} d\mathbf{r}''$$

$$\times \exp[i\mathbf{k}_1 \cdot (\mathbf{r}' - \mathbf{r}'')] \langle \Delta n_1(\mathbf{r}', t_l) \Delta n_1(\mathbf{r}'', t_i) \rangle, \qquad 1 \leqslant i \leqslant n$$

$$(5\text{-}29)$$

where the sum is extended over the $n!$ permutations of the set $t_1, t_2, \ldots, t_n$. In fact, terms containing expressions of the type

$$\int_{V_{sc}} d\mathbf{r}' \int_{V_{sc}} d\mathbf{r}'' \exp[i\mathbf{k}_1 \cdot (\mathbf{r}' + \mathbf{r}'')] \langle \Delta n_1(\mathbf{r}', t_l) \Delta n_1(\mathbf{r}'', t_i) \rangle$$

$$(5\text{-}30)$$

vanish due to the homogeneity assumption on the scattering system, so that only $n!$ terms survive.

From Eq. (5-29) it follows that

$$G^{(n)}(\mathbf{r}t_1, \ldots, \mathbf{r}t_n, \mathbf{r}t_n, \ldots, \mathbf{r}t_1) = \sum_{(p)} \sum_{\substack{l=1 \\ i \text{ all different}}}^{n} G^{(1)}(\mathbf{r}t_i, \mathbf{r}t_l)$$

$$(5\text{-}31)$$

with

$$G^{(1)}(\mathbf{r}t_i, \mathbf{r}t_l) = \gamma \int_{V_{sc}} d\mathbf{r}' \int_{V_{sc}} d\mathbf{r}'' \exp[i\mathbf{k}_1 \cdot (\mathbf{r}' - \mathbf{r}'')]$$

$$\times \langle \Delta n_1(\mathbf{r}', t_i) \Delta n_1(\mathbf{r}'', t_l) \rangle \qquad (5\text{-}32)$$

which agrees with the Gaussian factorization property introduced in Chapter IV [see Eq. (4-110)] (Bertolotti *et al.*, 1969).

It is worth observing that this conclusion possesses a wider validity, in the sense that it could as well be derived in the form of Eq. (4-110), which applies to more general space–time arguments. This obviously implies that experiments performed with more than one counter are also incapable of giving information not contained in $G^{(1)}$.

We note that the validity of the previous derivation holds independently of the value of the ratio V_c/V_{sc}, V_c being a characteristic correlation volume of the scattering medium, which plays an important role when Eq. (5-28) is not satisfied, as we shall see in the following.

Although the assumption of normal distribution for the density fluctuations is very familiar, it does not possess a general validity (i.e., the case of a fluid near its critical point, to be discussed). A general statement on the validity of the Gaussian properties of the scattered field can be made irrespective of the properties of the medium, only if the incident field is nonstochastic (a property we have assumed valid throughout this book) and the scattering volume V_{sc} contains a great number of correlation volumes V_c.

In this case we can take advantage of the same argument given by Glauber (1963b) [see also the *quantum central limit theorem* discussed by Klauder and Sudarshan (1968)] in order to prove the Gaussian behavior of radiation emitted by a chaotic source, by simply considering as independent sources each scattering element of volume V_c.[†] This ensures that the scattered field is Gaussian, which in turn implies that, for $V_{sc}/V_c \gg 1$, the medium always behaves as if it were Gaussian. We observe that, in the classical limit, one can as well resort to the *central limit theorem* in its generalized form (Rice, 1954), which states that the distribution of the sum of N statistically equivalent and independent random vectors $\mathbf{w}_1, \mathbf{w}_2, \ldots, \mathbf{w}_N$ approaches a normal

---

[†] Strictly speaking Glauber's argument applies if the *P*-representation relative to the field scattered by each volume V_c exists.

law (that is a joint Gaussian distribution for the components of $\mathbf{w} = \mathbf{w}_1 + \mathbf{w}_2 + \cdots + \mathbf{w}_N$) as $N \to \infty$. The Gaussian factorization property expressed by Eq. (4-110) is then achieved by considering as vectors $\mathbf{w}_i$ in $2n$ dimensions the quantities

$$[\operatorname{Re} \hat{E}_i(\mathbf{r}_1, t_1), \operatorname{Im} \hat{E}_i(\mathbf{r}_1, t_1), \ldots, \operatorname{Re} \hat{E}_i(\mathbf{r}_n, t_n), \operatorname{Im} \hat{E}_i(\mathbf{r}_n, t_n)] \qquad (5\text{-}33)$$

$$i = 1, 2, \ldots, N$$

where $\hat{E}_i$ is the field scattered by the ith correlation region, so that the limit $N \to \infty$ corresponds to $V_{sc}/V_c \to \infty$.

It is useful to give a direct derivation of the Gaussian factorization property, which hinges on the role that all-order correlations of density fluctuation play in the scattering process, since it allows us to give an estimate of the non-Gaussian term. To this end, let us begin by observing that the expression

$$\langle \Delta n_1 (\mathbf{r}_1, t) \Delta n_1 (\mathbf{r}'_1, t) \cdots \Delta n_1 (\mathbf{r}_n, t) \Delta n_1 (\mathbf{r}'_n, t) \rangle \qquad (5\text{-}34)$$

appearing on the right-hand side of Eq. (5-9) (we consider for the sake of simplicity and without loss of generality, the case $t_1 = t_2 = \cdots = t_n = t$; see footnote on p. 57) vanishes if the points $\mathbf{r}_1, \mathbf{r}'_1, \ldots, \mathbf{r}_n, \mathbf{r}'_n$ are all contained in zones of volume V_c such that at least one of these zones contains a single point. In fact, if the isolated point is for example $\mathbf{r}_1$, we have

$$\langle \Delta n_1 (\mathbf{r}_1, t) \Delta n_1 (\mathbf{r}'_1, t) \cdots \Delta n_1 (\mathbf{r}_n, t) \Delta n_1 (\mathbf{r}'_n, t) \rangle$$
$$= \langle \Delta n_1 (\mathbf{r}'_1, t) \cdots \Delta n_1 (\mathbf{r}_n, t) \Delta n_1 (\mathbf{r}'_n, t) \rangle \langle \Delta n_1 (\mathbf{r}_1, t) \rangle = 0$$
$$(5\text{-}35)$$

since $\langle \Delta n_1 \rangle = 0$, so that this point configuration does not contribute to the integral in Eq. (5-9). In general, the contributions to the $2n$-fold integration appearing in Eq. (5-9) are derived from various configurations, the simplest of which is such that the $2n$-space points $\mathbf{r}_1, \mathbf{r}'_1, \ldots, \mathbf{r}_n, \mathbf{r}'_n$ can be divided into n couples, each inside a different volume V_c. All the other configurations are such that the $2n$ points are distributed in less than n zones of volume V_c, for example, $2n - 4$ points in $n - 2$ zones of volume V_c and 4 in a single zone. The weight of the contribution to the integral of each type of configuration is, roughly speaking, proportional to the fraction of the total integration volume V_{sc}^{2n} pertinent to it. It is clear that the set of all configurations whose points are divided in couples covers an integration volume of the order of $(V_{\text{sc}} V_{\text{c}})^n$ as can easily be demonstrated by noting that, while one point of each couple moves through the volume V_{sc}, the other remains in a volume V_{c} around it. This set of configurations gives the more important contribution since the integration volume relative to any other set is not greater than

$$V_{\text{sc}}^{n-1} V_{\text{c}}^{n+1} = (V_{\text{sc}} V_{\text{c}})^n \frac{V_{\text{c}}}{V_{\text{sc}}} \tag{5-36}$$

(As an example, the second type of configuration considered above has at its disposal an integration volume just equal to $V_{\text{sc}}^{n-1} V_{\text{c}}^{n+1}$.) Therefore, whenever the expression in Eq. (5-34) appears in a multiple-volume integration as in Eq. (5-9), it can be assumed to factor in a Gaussian way as in Eq. (5-31), neglecting a term ε_n of order $V_{\text{c}}/V_{\text{sc}}$ in the result of the integration. This proves, in the same approximation, the validity of Eq. (5-31) for the scattered field. Along the same lines, we can immediately extend this proof to include the case of Eq. (4-110).

One has to note that the term ε_n is of the kind

$$\varepsilon_n = A_{\text{NG}} \frac{V_{\text{c}}}{V_{\text{sc}}} \tag{5-37}$$

where A_{NG} is a coefficient depending on the density fluctuation of the scattering system. The magnitude of A_{NG} increases according to the deviation of the statistics of the medium from a Gaussian one, a case in which it obviously vanishes.

The previous considerations provide the opportunity for separating the fluctuating scattering systems into two classes, with regard to the information contained in higher-order statistical measurements on the scattered radiation. The first class contains those media for which Eq. (5-28) holds true, so that the scattered field is Gaussian, independently of the scattering volume. In this case, no information can be gained by means of higher-order experiments, apart from the fact that the scattering medium is Gaussian. To the second class belong all those systems that do not satisfy Eq. (5-28). The result of higher-order experiments is different according to the value of the ratio V_c/V_{sc}. In fact, if $V_c/V_{sc} \ll 1$, the scattered field is Gaussian, while its behavior depends on the particular properties of the medium for V_c/V_{sc} of the order of unity. In this sense, the change of the statistical behavior of scattered radiation undergone by varying the scattering volume V_{sc} can yield a rough estimate of the size of V_c (provided this is large enough to be experimentally accessible). We shall deal in detail with this property in the next chapter in connection with the scattering of light by particles suspended in a turbulent fluid.

It is generally true that higher-order measurements performed on light scattered by a volume V_{sc} of the order of V_c can be used to investigate the structure of higher-order correlation functions of the medium, while for $V_{sc} \gg V_c$ it is often practically impossible to obtain information not contained in ordinary scattering experiments.

V.5   Statistical Behavior of the Scattered Field Observed with More Than One Detector

Up to now we have dealt with experiments involving one detector, thus limiting ourselves to consideration of the field scattered at a single angle, that is, field correlation functions evaluated at a single space point **r**. We have already noted at the end of Section V.3 that a measurement of correlation of intensity evaluated at different space points can be achieved by recording coincidences between counters placed at different positions. Although this kind of experiment does not seem very easily performed on scattered light at present, it is useful to examine briefly what kind of information one can expect to extract from them.

For the case in which the Gaussian nature of the scattered medium, expressed by Eq. (5-28), is a priori known, there is obviously no new information associated with this kind of measurement. On the contrary, an experiment performed with more than one counter is necessary to test possible departures of the properties of the scattering system from the Gaussian ones, whenever experimental conditions are such that $V_{sc} \gg V_c$ (Cantrell, 1968). In order to prove this fact let us consider an experiment in which the correlation of the intensities scattered at two different angles is measured (the generalization to the case in which more than two counters are employed is immediate). One has, according to Eqs. (5-4), (5-5), and (5-7),

$$G^{(2)}(\mathbf{r}t, \mathbf{r}'t', \mathbf{r}'t', \mathbf{r}t)$$

$$= \left(\frac{8\pi}{c}\right)^2 \langle I(\mathbf{r}, t) I(\mathbf{r}', t') \rangle$$

$$= \gamma^2 \int_{V_{sc}} d\mathbf{r}_1 \int_{V_{sc}} d\mathbf{r}'_1 \int_{V_{sc}} d\mathbf{r}_2 \int_{V_{sc}} d\mathbf{r}'_2$$

$$\times \exp[i\mathbf{k}_1 \cdot (\mathbf{r}_1 - \mathbf{r}'_1) + i\mathbf{k}'_1 \cdot (\mathbf{r}_2 - \mathbf{r}'_2)]$$

$$\times \langle \Delta n_1(\mathbf{r}_1, t) \Delta n_1(\mathbf{r}'_1, t) \Delta n_1(\mathbf{r}_2, t') \Delta n_1(\mathbf{r}'_2, t') \rangle$$

$$\tag{5-38}$$

where

$$\mathbf{k}_1 = \mathbf{k}_0 - k_0 \mathbf{r}_1 / r_1$$
$$\mathbf{k}'_1 = \mathbf{k}_0 - k_0 \mathbf{r}'_1 / r'_1$$

$$\tag{5-39}$$

If one introduces the non-Gaussian contribution

$$h(\mathbf{r}_1 t, \mathbf{r}'_1 t, \mathbf{r}_2 t', \mathbf{r}'_2 t')$$

to the fourth-order density correlation function

$$\langle \Delta n_1(\mathbf{r}_1, t)\, \Delta n_1(\mathbf{r}_1', t)\, \Delta n_1(\mathbf{r}_2, t')\, \Delta n_1(\mathbf{r}_2', t') \rangle$$

$$= \langle \Delta n_1(\mathbf{r}_1, t)\, \Delta n_1(\mathbf{r}_1', t) \rangle \langle \Delta n_1(\mathbf{r}_2, t')\, \Delta n_1(\mathbf{r}_2', t') \rangle$$

$$+ \langle \Delta n_1(\mathbf{r}_1, t)\, \Delta n_1(\mathbf{r}_2, t') \rangle \langle \Delta n_1(\mathbf{r}_1', t)\, \Delta n_1(\mathbf{r}_2', t') \rangle$$

$$+ \langle \Delta n_1(\mathbf{r}_1, t)\, \Delta n_1(\mathbf{r}_2', t') \rangle \langle \Delta n_1(\mathbf{r}_1', t)\, \Delta n_1(\mathbf{r}_2, t') \rangle$$

$$+ h(\mathbf{r}_1\, t, \mathbf{r}_1'\, t, \mathbf{r}_2\, t', \mathbf{r}_2'\, t') \tag{5-40}$$

Eq. (5-38) can be rewritten as

$$G^{(2)}(\mathbf{r}t, \mathbf{r}'t', \mathbf{r}'t', \mathbf{r}t)$$

$$= (8\pi/c)^2 \langle I(\mathbf{r}, t) \rangle \langle I(\mathbf{r}', t') \rangle$$

$$+ \gamma^2 \int_{V_{sc}} d\mathbf{r}_1 \int_{V_{sc}} d\mathbf{r}_1' \int_{V_{sc}} d\mathbf{r}_2 \int_{V_{sc}} d\mathbf{r}_2'$$

$$\times \exp\left[i\mathbf{k}_1 \cdot (\mathbf{r}_1 - \mathbf{r}_1') + i\mathbf{k}_1' \cdot (\mathbf{r}_2 - \mathbf{r}_2') \right] h(\mathbf{r}_1\, t, \mathbf{r}_1'\, t, \mathbf{r}_2\, t', \mathbf{r}_2'\, t')$$

$$\tag{5-41}$$

the second term on the right-hand side being of the order of V_c/V_{sc} with respect to the first term, according to the usual considerations.

In deriving Eq. (5-41) we have taken into account the fact that the second and third terms on the right-hand side of Eq. (5-40) do not contribute to the integral in Eq. (5-38) due to the homogeneity of the system, which has as a consequence the presence of oscillating factors of the kind

$$\int_{V_{sc}} \exp\left[i(\mathbf{k}_1 - \mathbf{k}_1') \cdot \mathbf{R} \right] d\mathbf{R} \tag{5-42}$$

vanishing for $\mathbf{k}_1$ appreciably different from $\mathbf{k}_1'$. Equation (5-41) differs from the Gaussian case, in which no correlation between the intensities is present, for the presence of a term containing h. Although this term is of the order of V_c/V_{sc} with respect to the Gaussian term, it can be observed experimentally since it depends on time difference $t' - t$, while the Gaussian contribution is time-independent due to stationarity. To this end, it is sufficient, for example, to perform the Fourier time transform of the intensity correlation with respect to the temporal argument $t' - t$. The Gaussian term will contribute only at the frequency $\omega = 0$, while for other frequencies one observes only the effect of h (Cantrell, 1968).

V.6 Scattering by a Stable Plasma

As a particular example of a physical system for which the assumption of normal distribution of density fluctuations [see Eq. (5-28)] is actually verified, we consider in this section a plasma at thermal equilibrium. Its dynamic behavior could be investigated by starting from the BBGKY hierarchy introduced in Chapter II. We adopt here the formally different approach based on the use of an equation that directly yields the evolution of the microscopic density (Klimontovich, 1967).

Let us consider a plasma consisting of N electrons and N ions, resulting from the ionization of N atoms of a given substance. The microscopic density v_1 in the N-dimensional phase space is given by [see Eq. (2-19)]

$$v_1^{(j)}(\mathbf{r}, \mathbf{v}, t) = \sum_{\alpha = 1}^{N} \delta [\mathbf{r} - \mathbf{r}_\alpha^{(j)}(t)] \, \delta [\mathbf{v} - \mathbf{v}_\alpha^{(j)}(t)] \qquad (5\text{-}43)$$

where the superscript j specifies the type of particles ($j = e$ for electrons and $j = i$ for ions). The equation of motion for $v_1^{(j)}$ is

$$\frac{\partial}{\partial t} v_1^{(j)} = \sum_{\alpha = 1}^{N} \left[\frac{d\mathbf{r}_\alpha^{(j)}}{dt} \cdot \frac{\partial v_1^{(j)}}{\partial \mathbf{r}_\alpha^{(j)}} + \frac{d\mathbf{v}_\alpha^{(j)}}{dt} \cdot \frac{\partial v_1^{(j)}}{\partial \mathbf{v}_\alpha^{(j)}} \right] \qquad (5\text{-}44)$$

where, in the absence of external forces,

$$\frac{d\mathbf{r}_\alpha^{(j)}}{dt} = \mathbf{v}_\alpha^{(j)}$$

(5-45)

$$m_j \frac{d\mathbf{v}_\alpha^{(j)}}{dt} = -\frac{\partial}{\partial \mathbf{r}_\alpha^{(j)}} \sum_{j'} \sum_{\beta=1}^{N} V_{jj'}(\mathbf{r}_\beta^{(j')} - \mathbf{r}_\alpha^{(j)})$$

with $V_{jj'}$ the interaction potential between a particle of type j' and a particle of type j with mass m_j and charge e_j.

By introducing Eqs. (5-45) in Eq. (5-44) we obtain, with the help of simple algebra, the *microscopic Vlasov equation*

$$\frac{\partial}{\partial t} v_1^{(j)} + \mathbf{v} \cdot \frac{\partial}{\partial \mathbf{r}} v_1^{(j)} + \frac{e_j}{m_j} \mathbf{E}_m(\mathbf{r}, t) \cdot \frac{\partial v_1^{(j)}}{\partial \mathbf{v}} = 0 \qquad (5\text{-}46)$$

where the relation

$$e_j \mathbf{E}_m(\mathbf{r}, t) = -\sum_{j'} \int \frac{\partial}{\partial \mathbf{r}} V_{jj'}(\mathbf{r} - \mathbf{r}') v_1^{(j')}(\mathbf{r}', \mathbf{v}', t) \, d\mathbf{r}' \, d\mathbf{v}' \qquad (5\text{-}47)$$

defines the fluctuating microscopic electric field $\mathbf{E}_m(\mathbf{r}, t)$.

It is worth mentioning, as we have already briefly observed in Section II.4, that one can immediately deduce the first equation of the BBGKY hierarchy by taking the ensemble average of Eq. (5-46). Subtracting this averaged equation from Eq. (5-46), we obtain

$$\frac{\partial}{\partial t} \Delta v_1^{(j)}(\mathbf{r}, \mathbf{v}, t) + \mathbf{v} \cdot \frac{\partial}{\partial \mathbf{r}} \Delta v_1^{(j)}(\mathbf{r}, \mathbf{v}, t) + \frac{e_j}{m_j} \frac{\partial}{\partial \mathbf{v}} \cdot [\mathbf{E}_m v_1^{(j)} - \langle \mathbf{E}_m v_1^{(j)} \rangle]$$

$$= 0 \qquad (5\text{-}48)$$

where

$$\Delta v_1^{(j)}(\mathbf{r}, \mathbf{v}, t) = v_1^{(j)}(\mathbf{r}, \mathbf{v}, t) - \langle v_1^{(j)}(\mathbf{r}, \mathbf{v}, t) \rangle \qquad (5\text{-}49)$$

Once Eq. (5-48) is solved, we can evaluate the quantity

$$\Delta n_1^{(j)}(\mathbf{r}, t) = \int \Delta v_1^{(j)}(\mathbf{r}, \mathbf{v}, t) \, d\mathbf{v} \qquad (5\text{-}50)$$

which allows us to determine in a direct way the correlation functions $\mathscr{G}_n(\mathbf{r}_1, \dots, \mathbf{r}_n, t_1, \dots, t_n)$ relevant to scattering [see Eq. (2-36)] with the help of suitable ensemble-averaging operations. Equation (5-48) can be cast in a closed form under suitable approximations. In fact, after writing

$$\mathbf{E}_m(\mathbf{r}, t) = \langle \mathbf{E}_m \rangle + \Delta \mathbf{E}_m(\mathbf{r}, t) \qquad (5\text{-}51)$$

with

$$\Delta \mathbf{E}_m(\mathbf{r}, t) = -\frac{1}{e_j} \sum_{j'} \int \frac{\partial}{\partial \mathbf{r}} V_{jj'}(\mathbf{r} - \mathbf{r}') \Delta v_1^{(j')}(\mathbf{r}', \mathbf{v}', t) \, d\mathbf{r}' \, d\mathbf{v}' \qquad (5\text{-}52)$$

Eq. (5-48) reduces to

$$\frac{\partial}{\partial t} \Delta v_1^{(j)}(\mathbf{r}, \mathbf{v}, t) + \mathbf{v} \cdot \frac{\partial}{\partial \mathbf{r}} \Delta v_1^{(j)}(\mathbf{r}, \mathbf{v}, t) + \frac{e_j}{m_j} \frac{\partial}{\partial \mathbf{v}} \cdot [\Delta \mathbf{E}_m \Delta v_1^{(j)}$$
$$- \langle \Delta \mathbf{E}_m \Delta v_1^{(j)} \rangle + \langle \mathbf{E}_m \rangle \Delta v_1^{(j)} + \Delta \mathbf{E}_m \langle v_1^{(j)} \rangle] = 0$$
$$\qquad (5\text{-}53)$$

which can be linearized in the form

$$\frac{\partial}{\partial t} \Delta v_1^{(j)} + \mathbf{v} \cdot \frac{\partial}{\partial \mathbf{r}} \Delta v_1^{(j)} + \frac{e_j}{m_j} \Delta \mathbf{E}_m \cdot \frac{\partial}{\partial \mathbf{v}} \langle v_1^{(j)} \rangle = 0 \qquad (5\text{-}54)$$

having taken advantage of the fact that for a homogeneous plasma $\langle \mathbf{E}_m \rangle = 0$. Neglecting the quadratic term in Δv_1 implies the weakness of the fluctuations, a circumstance verified in particular for the case of thermal equilibrium.

The system of Eqs. (5-52) and (5-54) leads to the evaluation of the Fourier transform in space and time of $\Delta v_1^{(j)}$ once $\langle v_1^{(j)} \rangle$ is assumed to be a known quantity. In effect, $\langle v_1 \rangle$ coincides with NF_1, F_1 being the one-particle distribution function [see Eq.

(2-20)] and takes, for example, the well-known Maxwellian form in the case of a plasma at thermal equilibrium. After using some algebra (see, for example, Fidone *et al.*, 1963), we arrive at the following expression for the Fourier transform of the electron density fluctuation:

$$\Delta \tilde{n}_1^{(e)}(\mathbf{k}, \omega) = B^{(e)}(\mathbf{k}, \omega) \tilde{n}_L^{(e)}(\mathbf{k}, \omega) + B^{(i)}(\mathbf{k}, \omega) \tilde{n}_L^{(i)}(\mathbf{k}, \omega)$$

$$(5\text{-}55)$$

where

$$B^{(e)} = \frac{1 - R^{(i)}(\mathbf{k}, \omega)}{1 - R^{(e)}(\mathbf{k}, \omega) - R^{(i)}(\mathbf{k}, \omega)}, \qquad B^{(i)} = 1 - B^{(e)}$$

$$(5\text{-}56)$$

and

$$R^{(j)}(\mathbf{k}, \omega) = \frac{4\pi e^2}{m_j k^2} \lim_{\varepsilon \to 0} \int \frac{\mathbf{k} \cdot \partial \langle v_1^{(j)} \rangle / \partial \mathbf{v}}{\mathbf{k} \cdot \mathbf{v} - \omega - i\varepsilon} \, d\mathbf{v} \qquad (5\text{-}57)$$

with $\tilde{n}_L^{(j)}(\mathbf{k}, \omega)$ the Fourier transform of the microscopic density evaluated along a free trajectory, that is,

$$\tilde{n}_L^{(j)}(\mathbf{k}, \omega) = \frac{1}{2\pi} \int_{-\infty}^{+\infty} dt \int d\mathbf{r} \exp(-i\mathbf{k} \cdot \mathbf{r} + i\omega t) \sum_{\alpha = 1}^{N} \delta[\mathbf{r} - \mathbf{r}_{\alpha 0}^{(j)} - \mathbf{v}_{\alpha 0}^{(j)} t]$$

$$= \sum_{\alpha = 1}^{N} \exp(i\mathbf{k} \cdot \mathbf{r}_{\alpha 0}^{(j)}) \delta(\omega - \mathbf{k} \cdot \mathbf{v}_{\alpha 0}^{(j)}) \qquad (5\text{-}58)$$

Equation (5-55) provides the basic relation by means of which the most general density correlation function can be evaluated. The statistical character of the problem is contained in the quantities $\tilde{n}_L^{(e)}(\mathbf{k}, \omega)$ and $\tilde{n}_L^{(i)}(\mathbf{k}, \omega)$, so that when performing the ensemble-averaging operation one has to evaluate expressions of the kind

$$\langle \tilde{n}_L^{(e)}(\mathbf{k}_1, \omega_1) \tilde{n}_L^{(e)}(\mathbf{k}_2, \omega_2) \cdots \tilde{n}_L^{(e)}(\mathbf{k}_j, \omega_j)$$

$$\times \tilde{n}_L^{(i)}(\mathbf{k}_{j+1}, \omega_{j+1}) \times \cdots \tilde{n}_L^{(i)}(\mathbf{k}_n, \omega_n) \rangle \qquad (5\text{-}59)$$

We shall limit ourselves to calculating the second- and fourth-order ensemble averages

$$\langle \Delta \tilde{n}_1^{(e)}(\mathbf{k}, t) \, \Delta \tilde{n}_1^{(e)}(-\mathbf{k}, t') \rangle, \quad \langle |\Delta \tilde{n}_1^{(e)}(\mathbf{k}, t)|^2 \, |\Delta \tilde{n}_1^{(e)}(\mathbf{k}, t')|^2 \rangle$$

(5-60)

relevant to the evaluation of the first- and second-order correlation functions of the scattered field. By Fourier inversion of Eq. (5-55) we obtain

$$\Delta \tilde{n}_1^{(e)}(\mathbf{k}, t) = \sum_{\alpha=1}^{N} B^{(e)}(\mathbf{k}, \mathbf{k} \cdot \mathbf{v}_{\alpha 0}^{(e)}) \exp [i\mathbf{k} \cdot (\mathbf{r}_{\alpha 0}^{(e)} - \mathbf{v}_{\alpha 0}^{(e)} t)]$$

$$+ \sum_{\alpha=1}^{N} B^{(i)}(\mathbf{k}, \mathbf{k} \cdot \mathbf{v}_{\alpha 0}^{(i)}) \exp [i\mathbf{k} \cdot (\mathbf{r}_{\alpha 0}^{(i)} - \mathbf{v}_{\alpha 0}^{(i)} t)]$$

(5-61)

having taken advantage of Eq. (5-58). In order to evaluate second- and fourth-order ensemble averages one must use the distribution functions F_1, F_2, F_3, and F_4 introduced in Chapter II, together with the analogous functions accounting for the correlation between particles of different species. By assuming as a first approximation the absence of any correlation [see Eq. (2-15)] that would be exactly verified for a perfect gas, the calculation can be easily performed by introducing the normalized one-particle velocity distributions $f^{(e)}(\mathbf{v})$ and $f^{(i)}(\mathbf{v})$. In practice, the previous assumption is valid for a nearly collisionless plasma, that is to the lowest significant order in the smallness parameter

$$\zeta = \frac{e^2}{K_B T \lambda_D} = \frac{1}{4\pi \langle n_1^{(e)} \rangle \lambda_D^3}$$

which is the ratio of the energy of interaction between two electrons separated by a distance of the *Debye length* $\lambda_D = (K_B T / 4\pi \langle n_1^{(e)} \rangle e^2)^{1/2}$ and their average kinetic energy $K_B T$ (see, for example, Oberman *et al.*, 1962; Fisher, 1964, Chapter IV).

Proceeding in this way one is able to derive,[†] for $kV_{sc}^{1/3} \gg 1$ (Crosignani *et al.*, 1968a)

$$\langle \Delta \tilde{n}_1^{(e)}(\mathbf{k}, t) \Delta \tilde{n}_1^{(e)}(-\mathbf{k}, t') \rangle$$

$$= N \int d\mathbf{v} \exp[-i\mathbf{k} \cdot \mathbf{v}(t-t')]$$

$$\times \{ |B^{(e)}(\mathbf{k}, \mathbf{k} \cdot \mathbf{v})|^2 f^{(e)}(\mathbf{v}) + |B^{(i)}(\mathbf{k}, \mathbf{k} \cdot \mathbf{v})|^2 f^{(i)}(\mathbf{v}) \}$$

$$(5\text{-}62)$$

and

$$\langle |\Delta \tilde{n}_1^{(e)}(\mathbf{k}, t)|^2 |\Delta \tilde{n}_1^{(e)}(\mathbf{k}, t')|^2 \rangle$$

$$= N \int d\mathbf{v} \, |B^{(e)}(\mathbf{k}, \mathbf{k} \cdot \mathbf{v})|^4 f^{(e)}(\mathbf{v}) + N^2 \int d\mathbf{v} \, d\mathbf{v}'$$

$$\times \{ 1 + \exp[-i\mathbf{k} \cdot (\mathbf{v} - \mathbf{v}')(t-t')] \} |B^{(e)}(\mathbf{k}, \mathbf{k} \cdot \mathbf{v})|^2 f^{(e)}(\mathbf{v})$$

$$\times \left[\frac{N(N-1)}{N^2} |B^{(e)}(\mathbf{k}, \mathbf{k} \cdot \mathbf{v}')|^2 f^{(e)}(\mathbf{v}') + |B^{(i)}(\mathbf{k}, \mathbf{k} \cdot \mathbf{v}')|^2 f^{(i)}(\mathbf{v}') \right]$$

+ the same terms with (i) changed to (e) and vice versa

$$(5\text{-}63)$$

Neglecting the terms linear in N with respect to those proportional to N^2, Eqs. (5-62) and 5-63) yield

$$\langle |\Delta \tilde{n}_1^{(e)}(\mathbf{k}, t)|^2 \, |\Delta \tilde{n}_1^{(e)}(\mathbf{k}, t')|^2 \rangle$$

$$= \langle |\Delta \tilde{n}_1^{(e)}(\mathbf{k}, t)|^2 \rangle + |\langle \Delta \tilde{n}_1^{(e)}(\mathbf{k}, t) \Delta \tilde{n}_1^{(e)}(-\mathbf{k}, t') \rangle|^2$$

$$(5\text{-}64)$$

which is consistent with the general Gaussian behavior expressed by Eq. (5-28).

[†] The evaluation of the density–density correlation function in connection with light scattering from a dense plasma, for which ζ is no longer small, has been examined by many authors (see, for example, the recent paper of Linnebur and Duderstadt, 1973). Unfortunately, no attention has been paid to higher-order correlation functions.

It is now necessary to evaluate the expression of the field scattered by a plasma, which can be immediately done by resorting to the general method introduced in Chapter I. If one assumes the frequency of the incident field ω_0 much greater than the characteristic electron *plasma frequency* ω_{p_e}, the motion $\mathbf{R}(t)$ induced on the single electron by the external field $\mathbf{E}(\mathbf{r}, t)$ is not coupled to its trajectory $\mathbf{r}(t)$ in the absence of $\mathbf{E}(\mathbf{r}, t)$. In fact, this trajectory is associated with the collective motion of the plasma and contains frequency of the order of $\omega_{p_e} = (4\pi \langle n_1^{(e)} \rangle e^2/m_e)^{1/2}$ or smaller. The induced dipole moment $\mathbf{P}$ is given by

$$\mathbf{P}(t) = e\mathbf{R}(t) \tag{5-65}$$

where $\mathbf{R}(t)$ obeys the equation

$$\ddot{\mathbf{R}}(t) = \frac{e}{m_e} \mathbf{E}_0 \cos\{\mathbf{k}_0 \cdot [\mathbf{r}(t) + \mathbf{R}(t)] - \omega_0 t\} \tag{5-66}$$

Equation (5-66) is readily integrated if the quantity $\mathbf{k}_0 \cdot [\mathbf{r}(t) + \mathbf{R}(t)]$ can be considered to undergo variations smaller than unity in a period $2\pi/\omega_0$. This circumstance is verified for the first term $\mathbf{k}_0 \cdot \mathbf{r}(t)$ in the nonrelativistic hypothesis, while $|\mathbf{k}_0 \cdot \mathbf{R}(t)| \ll 1$ if the ratio of the maximum excursion due to the incident field to the wavelength, which is given by $eE_0/m_e\omega_0 c$, is small.

Under this hypothesis, the oscillating part of the solution of Eq. (5-66), which is physically relevant to the evaluation of the scattered field, is

$$\mathbf{R}(t) = -\frac{e}{m_e\omega_0^2} \mathbf{E}_0 \cos\{\mathbf{k}_0 \cdot [\mathbf{r}(t) + \mathbf{R}(t)] - \omega_0 t\} \tag{5-67}$$

which corresponds to a polarizability $\alpha(\omega_0) = -e/m_e\omega_0^2$.

Once this quantity is introduced in Eq. (5-4), the singly scattered field is given by

$$\mathbf{E}_{sc}(\mathbf{r}, t) = \frac{r_0}{r} \boldsymbol{\eta} \times (\boldsymbol{\eta} \times \mathbf{E}_0) \exp\left[-i\omega_0(t-(r/c))\right] \tilde{n}_1^{(e)}\left(-\mathbf{k}_1, t-\frac{r}{c}\right) \tag{5-68}$$

(we neglect, because of the ion's large mass, the ionic contribution), where $r_0 = e^2/m_e c^2$ is the *classical electron radius*.

By taking advantage of Eq. (5-64) we then have

$$G^{(2)}(\mathbf{r}t_1, \mathbf{r}t_2, \mathbf{r}t_2, \mathbf{r}t_1)$$
$$= [G^{(1)}(\mathbf{r}t_1, \mathbf{r}t_1)]^2 + G^{(1)}(\mathbf{r}t_1, \mathbf{r}t_2)\, G^{(1)}(\mathbf{r}t_2, \mathbf{r}t_1)$$

$$(5\text{-}69)$$

which is consistent with Eq. (5-28). The validity of the more general Gaussian property expressed by Eq. (4-110) could be also easily shown by suitably generalizing Eq. (5-64).

We can conclude that the statistical Gaussian behavior of the field scattered by a stable plasma does not depend on the scattering volume, apart from the condition $N \gg 1$, which is almost always fulfilled in normal conditions. Thus higher-order measurements can only be regarded as a test of the linear theory responsible for Eq. (5-64). Once the validity of this theory has been confirmed, all information on the dynamic properties of the plasma is contained in ordinary cross-sectional measurements.

V.7 Intensity Correlation of Light Scattered by a Liquid Far from the Critical Point

The developments of the previous section are no longer valid when studying the structure of the correlation functions in a liquid (Fisher, 1964; Chapter V) since the analog of the smallness parameter ζ, which in the case of the plasma allows for a suitable procedure leading to explicit solutions such as the ones found for the second- and fourth-order correlation functions, does not exist. The macroscopic (hydrodynamical) approach considered in Section III.6 does not provide as well for the possibility of evaluating higher-order ensemble averages of density fluctuation products. One can, however, observe that, far from the critical point, the correlation length in a liquid is of the order of the range of the intermolecular potential (Fisher, 1964, Chapter V), which is so small that the scattering volume is always much greater than V_c. Thus the scattered field is, according to the results of Section V.4, a Gaussian variable. Therefore the knowledge of its spectrum is sufficient to determine higher-order correlation functions. This fact allows us conversely to introduce second-order measurements, as, for example, the conditional probability or the intensity-fluctuation spectrum defined in Section V.3, in order to determine the power spectrum. In particular, since the second-order Gaussian factorization property can be expressed as

$$\langle I(t)\,I(t+\tau)\rangle = \langle I(t)\rangle^2 + \left(\frac{c}{8\pi}\right)^2 |\langle \hat{E}(t)\,\hat{E}^*(t+\tau)\rangle|^2 \quad (5\text{-}70)$$

we have for the intensity-fluctuation spectrum [see Eq. (5-23)]

$$P(\omega) = \langle I(t)\rangle^2 \,\delta(\omega) + 4\pi^2 \int_{-\infty}^{+\infty} I(\omega')\,I(\omega'-\omega)\,d\omega' \quad (5\text{-}71)$$

where $I(\omega)$ is the ordinary power spectrum [see Eq. (2-48)].

To be more precise, a basic difference exists, for fixed V_c/V_{sc}, concerning the possible value that the factor A_{NG} in Eq. (5-37) can assume. It is clear that the value of A_{NG} has no practical influence on the statistics of the scattered light as long as $V_c/V_{sc} \to 0$, but it is characteristic of the dynamic behavior of the system. In fact, it is possible to show that A_{NG} vanishes for a liquid far from the critical point, as already seen for a stable plasma, which furnishes a case by far more interesting from an experimental point of view since the scattering volume can be easily made comparable with $V_c = \lambda_D^3$. We shall see that the vanishing of A_{NG} for a normal liquid is a direct consequence of the linearity of the set of equations describing the evolution of the system, as for the plasma case. This circumstance has a striking analogy with the correspondence between the linear character of the constitutive relations describing a macroscopic source and the Gaussian behavior of the emitted radiation

(Korenman, 1967). The loss of validity of the linear equations near the critical point will be shown conversely to give rise to a nonvanishing value of A_{NG}.

We start from the fluctuating hydrodynamical equations introduced in Chapter III. By taking the Fourier space transforms of Eq. (3-68) we write (see Tartaglia and Chen, 1973)

$$\frac{\partial \tilde{A}_\alpha(\mathbf{q}, t)}{\partial t} = -\sum_\beta L_{\alpha\beta}(\mathbf{q}) \, \tilde{A}_\beta(\mathbf{q}, t) + \tilde{f}_\alpha(\mathbf{q}, t) \qquad (5.72)$$

where $\tilde{A}_\alpha(\mathbf{q}, t)$ are the Fourier space transforms of the components of the column vector $A(\mathbf{r}, t) \equiv \{\rho(\mathbf{r}, t), \mathbf{v}(\mathbf{r}, t), T(\mathbf{r}, t)\}$. In Eq. (5-72), $L_{\alpha\beta}$ is a symmetric matrix and $\tilde{f}_\alpha(\mathbf{q}, t)$ are the Fourier transforms of the random forces, which vary in time much faster than the variables $\tilde{A}_\alpha(\mathbf{q}, t)$. In matrix form, Eq. (5-72) is

$$\frac{\partial \tilde{A}(\mathbf{q}, t)}{\partial t} = -L \tilde{A}(\mathbf{q}, t) + \tilde{f}(\mathbf{q}, t) \qquad (5-73)$$

with $L \equiv \{L_{\alpha\beta}\}$ and $\tilde{f} \equiv \{\tilde{f}_\alpha\}$.

Since a normal fluid deviates only slightly from thermo-dynamical equilibrium, the entropy corresponding to the small fluctuation $\tilde{A}(\mathbf{q}, t)^\dagger$ is given by (DeGroot and Mazur, 1962, Chapter VII)

$$S[\tilde{A}(\mathbf{q}, t)] = S_0 - \tfrac{1}{2} K_B \tilde{A}^+(\mathbf{q}, t) \, \varepsilon(\mathbf{q}) \, \tilde{A}(\mathbf{q}, t) \qquad (5-74)$$

where S_0 labels the equilibrium entropy, K_B the Boltzmann constant, and $\varepsilon(\mathbf{q})$ a positive-definite matrix. By comparing Eq. (5-74) with the relation connecting the probability density $w[\tilde{A}(\mathbf{q}, t)]$ and the entropy (*Boltzmann principle*)

$$S[\tilde{A}(\mathbf{q}, t)] = S_0 + K_B \ln w[\tilde{A}(\mathbf{q}, t)] \qquad (5-75)$$

---

† We observe that $A(\mathbf{r}, t) = A_0 + A'(\mathbf{r}, t)$, where A_0 is the equilibrium value independent of space and time, so that $\tilde{A}_0(\mathbf{q}, t) = 0$ and $\tilde{A}(\mathbf{q}, t)$ only contains the deviation $\tilde{A}'(\mathbf{q}, t)$ from equilibrium.

we obtain

$$w[\tilde{A}(\mathbf{q}, t)] = w_0 \exp\left[-\tfrac{1}{2}\tilde{A}^+(\mathbf{q}, t)\varepsilon(\mathbf{q})\,\tilde{A}(\mathbf{q}, t)\right] \qquad (5\text{-}76)$$

which means that the components of $\tilde{A}$ follow a joint Gaussian distribution (see Section V.4) if they are evaluated at the same time, so that in particular $\rho(\mathbf{q}, t)$ is a Gaussian variable.

In order to extend the preceding conclusion to more general temporal arguments, it is convenient to write a formal solution of Eq. (5-73) in terms of *Green's function* as

$$\tilde{A}(\mathbf{q}, t) = G(\mathbf{q}, t)\,\tilde{A}(\mathbf{q}, 0) + \int_0^t G(\mathbf{q}, t-s)\,\tilde{f}(\mathbf{q}, s)\,ds \qquad (5\text{-}77)$$

where $G(\mathbf{q}, t)$ is a matrix. Furthermore, the rapid time variation of $\tilde{f}$ allows us to consider it as uncorrelated with $\tilde{A}(\mathbf{q}, t)$, so that

$$\langle \tilde{A}_\alpha^*(\mathbf{q}, 0)\,\tilde{f}_\beta(\mathbf{q}, t)\rangle = \langle \tilde{A}_\alpha^*(\mathbf{q}, 0)\rangle \langle \tilde{f}_\beta(\mathbf{q}, t)\rangle = 0 \qquad (5\text{-}78)$$

and

$$\langle \tilde{f}_\alpha^*(\mathbf{q}, t)\,\tilde{f}_\beta(\mathbf{q}, t')\,\tilde{A}_\gamma^*(\mathbf{q}, 0)\,\tilde{A}_\gamma(\mathbf{q}, 0)\rangle$$
$$= \langle \tilde{f}_\alpha^*(\mathbf{q}, t)\,\tilde{f}_\beta(\mathbf{q}, t')\rangle \langle \tilde{A}_\gamma^*(\mathbf{q}, 0)\,\tilde{A}_\gamma(\mathbf{q}, 0)\rangle \qquad (5\text{-}79)$$

The factorization properties contained in Eqs. (5-76), (5-78), and (5-79), together with the dynamic evolution of $\tilde{A}(\mathbf{q}, t)$ expressed by Eq. (5-77), can be used in order to write fourth-order averages of the type

$$\langle \tilde{A}_\alpha^*(\mathbf{q}, 0)\,\tilde{A}_\beta(\mathbf{q}, 0)\,\tilde{A}_\gamma^*(\mathbf{q}, t)\,\tilde{A}_\delta(\mathbf{q}, t)\rangle$$

in terms of second-order averages. After using some algebra, we obtain in particular

$$\langle \tilde{\rho}_1^*(\mathbf{q}, 0)\,\tilde{\rho}_1(\mathbf{q}, 0)\,\tilde{\rho}_1^*(\mathbf{q}, t)\,\tilde{\rho}_1(\mathbf{q}, t)\rangle$$
$$= \langle |\tilde{\rho}_1(\mathbf{q}, 0)|^2\rangle + \langle \tilde{\rho}_1^*(\mathbf{q}, 0)\,\tilde{\rho}_1(\mathbf{q}, t)\rangle$$
$$\times \langle \tilde{\rho}_1(\mathbf{q}, 0)\,\tilde{\rho}_1^*(\mathbf{q}, t)\rangle \qquad (5\text{-}80)$$

Equation (5-80) is consistent with a normal distribution of the macroscopic density fluctuation ρ_1. On the other hand, since a direct relation exists between the scattered electric field and the Fourier space transform of the macroscopic density analogous to Eq. (5-4) (as can easily be seen by means of the macroscopic approach illustrated in Section III.2)

$$\hat{E}(\mathbf{r}, t) \propto \tilde{\rho}_1(-\mathbf{k}_1, t) \qquad (5\text{-}81)$$

we immediately obtain the Gaussian factorization property given by Eq. (5-69).

The most interesting aspect of the preceding derivation is that it is not necessary to make any assumption about the statistics of the random forces governing the process. This fact stresses the direct connection between the linear character of Eq. (5-73) and the Gaussian behavior of the density correlation at different times.

V.8 Intensity Correlation of Light Scattered by a Fluid near the Critical Point

The behavior of a fluid in a region near its critical point such that the relation $k_0 \xi \ll 1$ is not satisfied (see Section III.7) is characterized by the nonlinearity of the corresponding dynamic equations (Kadanoff and Swift, 1968; Kawasaki, 1971). This is related to the existence of mode–mode coupling between different modes, which is reponsible for the anomalies in various transport coefficients. Furthermore, the deviations from thermodynamical equilibrium are no longer small enough to allow for the approximation leading to the expansion for the entropy given by Eq. (5-74). Therefore, we do not expect a Gaussian distribution of $\tilde{\rho}_1(\mathbf{q}, t)$ in the critical region. This circumstance, together with the increase in the value of the correlation volume V_c suggests that we look for a deviation from the normal behavior in the statistics of light scattered by a fluid near the critical point.

While no analytical treatment has been performed in the general case, it is interesting to consider the equal time distribution of $\tilde{\rho}(\mathbf{q}, t)$ in such a way as to give an estimate of the non-Gaussian contribution (Korenman, 1970; Tartaglia and Chen, 1973). To this end, we observe that the correlation function $\langle \tilde{\rho}_1(\mathbf{k}_1, t) \tilde{\rho}_1(-\mathbf{k}_1, t) \rangle$ can be reasonably assumed continuous at $\mathbf{k}_1 = 0$ and slowly varying in the range of values of $\mathbf{k}_1$ relevant for a scattering experiment. This is due to the fact that the correlation length l_c, at which $\langle \rho_1(\mathbf{r}, t) \rho_1(0, t) \rangle$ is vanishing, does not practically exceed $1/2k_0$, whose inverse defines the upper limit of $\mathbf{k}_1$. A similar argument holds for higher-order correlation functions, and the determination of the statistics of $\tilde{\rho}_1(-\mathbf{k}_1, t)$ at $\mathbf{k}_1 = 0$ gives an insight into its behavior for more general values of $\mathbf{k}_1$ corresponding to the field scattered at nonvanishing angles.

Starting from the definition of Fourier space transform, one immediately obtains

$$\tilde{\rho}_1(0, t) = N - \langle N \rangle = \Delta N \qquad (5\text{-}82)$$

where N and $\langle N \rangle$ are, respectively, the number of particles contained in the scattering volume V_{sc} and its ensemble average, and ΔN is a statistical variable with zero average. We are interested in the deviation of the distribution function of this variable from a Gaussian variable.[†] To this end, we introduce the *grand partition function* of the system

$$Q = \text{Tr}[\exp(-\alpha N - H/K_B T)] \qquad (5\text{-}83)$$

where the symbol Tr indicates the operation of tracing over a complete set of orthogonal quantum states, $\alpha = \mu/K_B T$, where μ is the chemical potential and H represents the Hamiltonian.

[†] Of course, ΔN cannot be exactly Gaussian, since $\Delta N \geqslant -\langle N \rangle$. The validity of the following consideration is based on the assumption that $\langle N \rangle \gg 1$.

Then we have

$$-\left(\frac{\partial \ln Q}{\partial \alpha}\right)_T = \langle N \rangle \qquad (5\text{-}84)$$

$$\left(\frac{\partial^2 \ln Q}{\partial \alpha^2}\right)_T = \left(-\frac{\partial \langle N \rangle}{\partial \alpha}\right)_T = \langle (\Delta N)^2 \rangle \qquad (5\text{-}85)$$

$$\left(\frac{\partial^4 \ln Q}{\partial \alpha^4}\right)_T = \left(-\frac{\partial^3 \langle N \rangle}{\partial \alpha^3}\right)_T = \langle (\Delta N)^4 \rangle - 3\langle (\Delta N)^2 \rangle^2 \qquad (5\text{-}86)$$

the proof of Eqs. (5-84)–(5-86) hinging on Eq. (4-85), which allows us to express the statistical average quantum value of any operator in terms of the density matrix ρ. It is in the present case

$$\rho = \frac{\exp(-\alpha N - H/K_B T)}{\mathrm{Tr}[\exp(\alpha N - H/K_B T)]} \qquad (5\text{-}87)$$

where the term $-\alpha N$ in the exponential takes into account the possible fluctuation of particle number present in the volume V.

We now recall that the moments of a stochastic variable x, defined as

$$M(n) = \int_{-\infty}^{+\infty} (x - \langle x \rangle)^n f(x)\,dx \qquad (5\text{-}88)$$

assume, for a normal distribution $f(x)$, the form

$$M(2n) = (2n-1)!!\,[M(2)]^n \qquad (5\text{-}89)$$

valid for even order, while the odd-order moments vanish. If we specify $x = \Delta N$, it is clear that the value of the right-hand side of Eq. (5-86) is related to the deviation of the distribution of ΔN from a Gaussian one.

In order to estimate this deviation, we must use a suitable model of the equation of state for the fluid in the critical region (Korenman, 1970; Tartaglia and Chen, 1973), which determines the dependence of the equilibrium density $\rho_0 = \langle N \rangle / V_{sc}$ on μ and T. The procedure allows us to evaluate the non-Gaussian term NGT defined as

$$NGT = \frac{\langle (\Delta N)^4 \rangle - 3 \langle (\Delta N)^2 \rangle^2}{\langle (\Delta N)^2 \rangle^2} \qquad (5\text{-}90)$$

which turns out to be, as expected, of the type

$$NGT = A_{NG} V_{sc} / V_c \qquad (5\text{-}91)$$

A comparison between the results of Korenman (1970) and Tartaglia and Chen (1973) suggests a strong dependence of NGT on the particular fluid and on the value of $T - T_{cr}$, where T_{cr} is the critical temperature. In particular, Korenman claims that the detection of the non-Gaussian contribution in the scattered field is marginally possible for a simple fluid.

The possibility of applying the method described in Section V.5 (Cantrell, 1968) in order to extract information on the statistics of a fluid near the critical point from the non-Gaussian contribution to the correlation of intensities measured by means of two counters placed at different scattering angles has been considered by Swift (1973).

V.9 Self-Beating and Heterodyne Detections

It is convenient at this point to compare briefly the spectral measurement of light in terms of the power spectrum $I(\omega)$ and of the intensity spectrum $P(\omega)$. In connection with the developments of this chapter, it is clear that the two methods furnish different kinds of information if the electromagnetic radiation is not Gaussian. In the Gaussian case, while it is always possible to obtain $P(\omega)$ in terms of $I(\omega)$ [see Eq. (5-71)], the converse is not true. In fact, $P(\omega)$ does not depend on the phase of $\langle \hat{E}(t) \hat{E}^*(t + \tau) \rangle$, as is easily seen from Eq. (5-70).

In order to consider more explicitly the case of scattered light, let us write the scattered field in the form

$$\hat{E}(t) = \exp[i\psi(t) - i\omega_0 t]\, e(t) \qquad (5\text{-}92)$$

where $\psi(t)$ represents a fluctuating phase, the stochastic behavior of the scattering medium being contained in the (complex) quantity $e(t)$. Equation (5-92) is the generalization (Mandel, 1969) of Eq. (5-4) to the case in which the phase ψ of the incident field (of constant amplitude) is a slowly varying random function of time. It is obvious that the presence of ψ does not alter the value of $\langle I(t)\,I(t+\tau)\rangle$, while Eq. (5-70) is modified as

$$\langle I(t)\,I(t+\tau)\rangle = \langle I(t)\rangle^2 + \left(\frac{c}{8\pi}\right)^2 |\langle e(t)\,e^*(t+\tau)\rangle|^2 \quad (5\text{-}93)$$

Thus this kind of measurement (*intensity fluctuation spectroscopy* or *self-beating detection*) is not affected by the presence of a phase fluctuation in the incident field, although it does not furnish the phase of $\langle e(t)\,e^*(t+\tau)\rangle$.

The fluctuating phase of the incident field or, equivalently, its bandwidth cannot be conversely ignored when evaluating $\langle \hat{E}(t)\,\hat{E}^*(t+\tau)\rangle$, a quantity that is:

$$\langle \hat{E}(t)\,\hat{E}^*(t+\tau)\rangle = \langle \exp\{i[\psi(t) - \psi(t+\tau)]\}\rangle_F \langle e(t)\,e^*(t+\tau)\rangle_M$$

$$(5\text{-}94)$$

where the subscripts F and M indicate, respectively, the ensemble average over the light field and the material medium. Whenever the spectral width of the incident field is larger than that associated with $e(t)$, the behavior of $\langle \hat{E}(t)\,\hat{E}^*(t+\tau)\rangle$, and thus that of the optical spectrum $I(\omega)$ as given by Eq. (2-48), tends to be dominated by the fluctuations of $\psi(t)$ (Mandel, 1969).

There is, however, an experimental technique, *heterodyne detection*, that allows a determination of the optical spectrum avoiding the difficulties connected with the fluctuations of $\psi(t)$. This kind of measurement is basically second-order since the detector is illuminated simultaneously by the field under study $\hat{E}(t)$ and by a reference field $\hat{E}_0(t)$, and one observes the spectrum of the fluctuation of the instantaneous intensity $I(t)$

$$I(t) = \frac{c}{8\pi} |\hat{E}(t) + \hat{E}_0(t)|^2 \qquad (5\text{-}95)$$

If the relation $|\hat{E}_0(t)|^2 \gg |\hat{E}(t)|^2$ is satisfied, one finds in a stationary case that $\langle I(t) I(t+\tau) \rangle$ is given, apart from a time-independent term, by the expression

$$\langle I(t) I(t+\tau) \rangle = \left(\frac{c}{8\pi}\right)^2 \langle \hat{E}_0^*(t) \hat{E}_0(t+\tau) \hat{E}(t) \hat{E}^*(t+\tau) \rangle$$

$$+ \left(\frac{c}{8\pi}\right)^2 \langle \hat{E}_0(t) \hat{E}_0^*(t+\tau) \hat{E}^*(t) \hat{E}(t+\tau) \rangle$$

$$(5\text{-}96)$$

having neglected the term $\langle \hat{E}^*(t) \hat{E}(t) \hat{E}^*(t+\tau) \hat{E}(t+\tau) \rangle$.

In a scattering experiment, one uses the incident field as reference field so that the contributions of the fluctuating phase cancel on the right-hand side of Eq. (5-96), which reduces to

$$\langle I(t) I(t+\tau) \rangle = \frac{c}{8\pi} I_0 \langle e^*(t) e(t+\tau) \rangle + \text{c.c.} \qquad (5\text{-}97)$$

where I_0 is the intensity of the incident field. One then has [see Eq. (5-23)]

$$P(\omega) = I_0 [I_e(\omega) + I_e(-\omega)] \qquad (5\text{-}98)$$

having defined

$$I_e(\omega) = \frac{c}{16\pi^2} \int_{-\infty}^{+\infty} \langle e^*(t) e(t+\tau) \rangle e^{i\omega\tau} \, d\tau \qquad (5\text{-}99)$$

We observe that $I_e(\omega)$ is, in general, different from zero also for $\omega < 0$, since $e(t)$ is not an analytical signal. In particular, $I_e(\omega) = I_e(-\omega)$ whenever $\langle e^*(t) e(t+\tau) \rangle$ is a real quantity, which in turn implies

$$P(\omega) = 2I_0 I_e(\omega) \qquad (5\text{-}100)$$

In order to see when Eq. (5-100) holds true, we note that, according to the general expression of the scattered field, one has

$$\langle e^*(t)\,e(t+\tau)\rangle = g\langle \tilde{n}_1(\mathbf{k}_1,t)\,\tilde{n}_1(-\mathbf{k}_1,t+\tau)\rangle \quad (5\text{-}101)$$

where g is a positive constant. On the other hand,

$$\langle \tilde{n}_1(\mathbf{k}_1,t)\,\tilde{n}_1(-\mathbf{k}_1,t+\tau)\rangle = \int_{V_{sc}} d\mathbf{r} \int_{V_{sc}} d\mathbf{r}'\, \exp(i\mathbf{k}_1\cdot\mathbf{r}' - i\mathbf{k}_1\cdot\mathbf{r})$$

$$\times \langle n_1(\mathbf{r},t)\,n_1(\mathbf{r}',t+\tau)\rangle \quad (5\text{-}102)$$

which is a real quantity if the scattering system is isotropic, since this implies

$$\langle n_1(\mathbf{r},t)\,n_1(\mathbf{r}',t+\tau)\rangle = \langle n_1(\mathbf{r},t+\tau)\,n_1(\mathbf{r}',t)\rangle \quad (5\text{-}103)$$

Thus the isotropy assumption yields Eq. (5-100). This assumption holds true for many practical situations in which the average velocity of each scattering center is zero. There are, however, cases in which the presence of an average velocity $\mathbf{v}_0$ introduces a preferential direction. If the system is isotropic in a reference frame moving with velocity $\mathbf{v}_0$, one has

$$\langle \tilde{n}_1(\mathbf{k}_1,t)\,\tilde{n}_1(-\mathbf{k}_1,t+\tau)\rangle$$

$$= \exp(i\mathbf{k}_1\cdot\mathbf{v}_0 t)\langle \tilde{n}_1(\mathbf{k}_1,t)\,\tilde{n}_1(-\mathbf{k}_1,t+\tau)\rangle_{\mathbf{v}_0} \quad (5\text{-}104)$$

where the symbol $\langle\cdots\rangle_{\mathbf{v}_0}$ implies that the quantity in the angle brackets is evaluated in the moving frame. This equation can be immediately proved by using Eq. (2-52) and the Galilean composition law of velocities. By means of Eq. (5-104) and (5-99) one obtains

$$P(\omega) = I_0[I_{e,\mathbf{v}_0}(\omega+\Omega) + I_{e,\mathbf{v}_0}(\omega-\Omega)] \quad (5\text{-}105)$$

with $\Omega = \mathbf{k}_1\cdot\mathbf{v}_0$, and

$$I_{e,\mathbf{v}_0}(\omega) = \frac{c}{16\pi^2}\int_{-\infty}^{+\infty}\langle e^*(t)\,e(t+\tau)\rangle_{\mathbf{v}_0}e^{i\omega\tau}\,d\tau \quad (5\text{-}106)$$

Equation (5-105) shows that the heterodyne detection scheme

is sensitive to the presence of a term $\exp[i\mathbf{k}_1 \cdot \mathbf{v}_0 t]$ in Eq. (5-104), while this would completely disappear in self-beating detection [see Eq. (5-93)].

Finally, it is worth emphasizing the fact that heterodyne detection allows a measurement of the power spectrum of the field under study independently of its statistics.

The self-beating method has been developed in the optical range by Ford and Benedek (1965), while the heterodyne technique was first applied in connection with scattering experiments using the incident laser light as reference field by Cummins *et al.* (1964) and Yeh and Cummins (1964). For a detailed theoretical and experimental comparison of the two methods see, for example the work of Cummins and Swinney (1970) and Benedek (1969).

VI

Statistical Properties of Light Scattered by Small Particles

VI.1 Introduction

The scattered field considered in the preceding chapters is composed of contributions pertaining to elementary scattering units (atoms, molecules, or electrons), whose trajectories are independent parameters necessary to determine the properties of the scattered radiation. One can treat, along the same lines, scattering by macromolecules and other material particles whose dimensions exceed the molecular ones, considering them as single-scattering centers. In the first case the molecules are separated by empty space, the material particles being usually embedded in a medium that supports them without directly contributing to scattering. The trajectories of the molecules are essentially determined by their mutual interactions, while the motion of the particles depends, in general, only on the surrounding medium. This in turn implies that the information contained in the scattered field concerns, respectively, the dynamics of the molecular system and of the medium in which the particles are embedded. It is precisely this last circumstance that, together with the large scattering efficiency associated with the particle dimensions, allows one to investigate certain properties of some material systems by examining the light scattered by impurities that have been added to them. We shall consider in this chapter some applications of this technique, particularly to cases in which a direct investigation of the system by means of scattering would be very difficult.

Light scattering can be important for studying the properties of the particle system itself, in addition to those of the supporting medium. This is particularly interesting, for example, when the particles are self-propelled microorganisms that have their own dynamical behavior independent of the surrounding fluid.

Light scattering from small particles has been already the object of experimental investigations concerning higher-order statistical properties (for an extensive review, see, for example, Cummins, 1974). The existence of relevant situations in which second-order measurements have shown a non-Gaussian behavior of the scattered field, thus furnishing information not obtainable with ordinary methods, makes the investigation of this kind of scattering particularly interesting (Di Porto *et al.*, 1969; Bertolotti *et al.*, 1971; Schaefer and Berne, 1972).

This chapter is concerned, in particular, with the evaluation of the first-order correlation function of the field, thus yielding the expression of the optical spectrum $I(\omega)$. Significant spectral measurements in this area have become possible only in recent years, the main difficulty arising from the small frequency broadening due to the slow velocities associated with the inertia of the particles. In this respect, we observe that the scattering from macromolecules and small particles is usually referred to as *quasi-elastic*.

VI.2 The Quasi-Elastically Scattered Field

The study of the angular distribution of light scattered by small particles (considered at rest) has been the object of a large amount of research since the classical investigations on the color of the sky initiated by Tyndall and Rayleigh. The scattered field has been obtained in a vast number of situations concerning the size, shape, and refractive index of the scattering centers, one of the main applications of this kind of scattering being the determination of the size distribution of colloidal particles. An extensive review of the argument is given in the classical texts of van de Hulst (1957), Born and Wolf (1970), and in the work of Kerker (1969).

The advent of the laser and of sophisticated methods of measurements, such as the one provided by self-beating and heterodyne techniques, has given relevance to the problem of determining the frequency spectrum of the scattered field. It is to be expected that a frequency broadening takes place when the scattering particles are no longer in a fixed position but undergo translational and rotational motions giving rise to Doppler shifts. (We do not consider in the following more complicated motions associated with other degrees of freedom, thus limiting ourselves to the treatment of rigid scatterers.) This problem was first investigated by Pecora (1964), who studied, from a microscopic point of view, scattering by solutions of polymer molecules.

The evaluation of the scattered field is greatly simplified by the so-called *Rayleigh–Gans hypothesis*, which says, from a physical point of view, that the scatterer introduces a small optical perturbation in the host medium and that each element of the scatterer itself sees the incident field practically unperturbed. Mathematically, these conditions are (van de Hulst, 1957, Chapter VII)

$$|n - n_0| \ll 1 \tag{6-1}$$

and

$$k_0 d |n - n_0| \ll 1 \qquad (6\text{-}2)$$

where n_0 and n represent, respectively, the refractive indices of the medium and of the particle in suspension and d is the order of magnitude of the particle's linear dimension. Equation (6-2) expresses the fact that the phase of the field does not appreciably vary due to the presence of the scatterer; the same condition on the amplitude can be shown to hold by comparing the overall scattered radiation with the radiation falling on the particle (van de Hulst, 1957, Chapter VII). We observe that Eq. (6-2) does not necessarily imply the condition $k_0 d \ll 1$, which pertains to Tyndall scattering.

We assume that the scatterer is composed of optically isotropic molecules so that its dielectric constant ε is a scalar quantity. We can now apply to the present situation, that is, a single scatterer acted upon by an external field, the theory developed in Section III.2 on scattering by dielectric constant fluctuations. To this end, it is sufficient to consider the dielectric constant fluctuation $\varepsilon_1(\mathbf{r}, t)$ as due to the presence of the particle, so that

$$\varepsilon_1(\mathbf{r}, t) = \varepsilon(\mathbf{r}, t) - \varepsilon_0 \qquad (6\text{-}3)$$

where ε_0 is the dielectric constant of the supporting medium. In fact, as already observed, Eqs. (6-1) and (6-2) imply that the incident field travels unperturbed, which is the condition underlying the developments of Section III.2. Thus we can use Eqs. (3-14) and (3-17), which supply the scattered field, in the form

$$\hat{\mathbf{E}}_1(\mathbf{r}, t) = \frac{-k_0^2 \, \boldsymbol{\eta} \times (\boldsymbol{\eta} \times \mathbf{E}_0)}{4\pi r} \int_V d\mathbf{r}' \, \varepsilon_1 \left(\mathbf{r}_1, t - \frac{n_0 r}{c} + \frac{n_0}{c} \frac{\mathbf{r} \cdot \mathbf{r}'}{r} \right)$$

$$\times \exp\left[i\mathbf{k}_1 \cdot \mathbf{r}' - i\omega_0 \left(\frac{t - n_0 r}{c} \right) \right] \qquad (6\text{-}4)$$

where V is the particle volume, $k_0 = n_0 \omega_0 / c$, and $\mathbf{k}_1 = \mathbf{k}_0 - k_0 \boldsymbol{\eta}$.

By taking Eqs. (6-1) and (6-3), into account Eq. (6-4) can be rewritten as

$$\hat{\mathbf{E}}_1(\mathbf{r},t) = - \frac{k_0^2 \boldsymbol{\eta} \times (\boldsymbol{\eta} \times \mathbf{E}_0) \exp(in_0\, r\omega_0/c)}{4\pi r} 2n_0 e^{-i\omega t} \int_V d\mathbf{r}'$$

$$\times \exp(i\mathbf{k}_1 \cdot \mathbf{r}')[n(\mathbf{r}',t) - n_0] \tag{6-5}$$

where the temporal arguments have been simplified, as usual, since we are interested in evaluating stationary ensemble averages that involve correlation times much larger than $V_{sc}^{1/3}/c$, V_{sc} being the scattering volume. If one assumes the scatterer homogeneous, that is, $n(\mathbf{r},t) = n$ over the particle volume, Eq. (6-5) takes the form

$$\hat{\mathbf{E}}_1(\mathbf{r},t) = - \frac{k_0^2 \boldsymbol{\eta} \times (\boldsymbol{\eta} \times \mathbf{E}_0) \exp(in_0\, r\omega_0/c)}{4\pi r} 2n_0(n-n_0) e^{-i\omega t}$$

$$\times \int_{V_t} \exp(i\mathbf{k}_1 \cdot \mathbf{r}')\, d\mathbf{r}' \tag{6-6}$$

where V_t changes its position and orientation with time, while its magnitude remains unchanged. Equation (6-6) can be similarly rewritten as

$$\hat{\mathbf{E}}_1(\mathbf{r},t) = - \frac{k_0^2 \boldsymbol{\eta} \times (\boldsymbol{\eta} \times \mathbf{E}_0) \exp(in_0\, r\omega_0/c)}{4\pi r} 2n_0(n-n_0)$$

$$\times \exp[i\mathbf{k}_1 \cdot \mathbf{r}_i(t) - i\omega_0 t] \int_{\tilde{V}_t} e^{i\mathbf{k}_1 \cdot \boldsymbol{\rho}}\, d\boldsymbol{\rho} \tag{6-7}$$

where $\mathbf{r}_i(t)$ represents the trajectory of a point of the particle and $\tilde{V}_t$ is the volume described by the vector $\boldsymbol{\rho} = \mathbf{r}' - \mathbf{r}_i(t)$, when $\mathbf{r}'$ varies over V_t (see Fig. 6.1). Although $\mathbf{r}_i(t)$ can be choosen in an arbitrary way, its most convenient characterization depends on the specific problem.

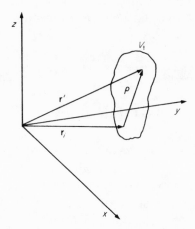

Fig. 6.1   *Geometrical relation for evaluating $R_t(\mathbf{\eta})$.*

The field scattered by the single particle is in general of the type

$$\hat{\mathbf{E}}_1(\mathbf{r}, t) = \mathbf{B} R_t(\mathbf{\eta}) \exp[i\mathbf{k}_1 \cdot \mathbf{r}_i(t) - i\omega_0 t] \qquad (6\text{-}8)$$

with

$$\mathbf{B} = -\frac{k_0^2 \mathbf{\eta} \times (\mathbf{\eta} \times \mathbf{E}_0)}{4\pi r} \exp\left[\frac{in_0 \omega_0 r}{c}\right] 2n_0(n - n_0) \qquad (6\text{-}9)$$

where the temporal dependence of

$$R_t(\mathbf{\eta}) = \int_{\tilde{V}_t} e^{i\mathbf{k}_1 \cdot \mathbf{\rho}} \, d\mathbf{\rho} \qquad (6\text{-}10)$$

is connected with the orientational motion around the point $\mathbf{r}_i(t)$.

In particular, for Tyndall scattering, one has

$$R_t(\mathbf{\eta}) = V \qquad (6\text{-}11)$$

and $\mathbf{r}_i(t)$ is the well-defined position of the particle. For a spherical scatterer, choosing $\mathbf{r}_i(t)$ as the trajectory of the center, one still has a time-independent value of $R_t(\eta)$ given by

$$R_t(\eta) = 4\pi \frac{a^2}{k_1} \left\{ \frac{\sin k_1 a}{(k_1 a)^2} - \frac{\cos k_1 a}{k_1 a} \right\}$$

$$= 4\pi \frac{\sin k_1 a}{k_1^3} - \frac{4\pi a}{k_1^2} \cos k_1 a$$

$$= \left(\frac{2\pi a}{k_1} \right)^{3/2} J_{3/2}(k_1 a) \tag{6-12}$$

where $J_{3/2}$ is the Bessel function of order $3/2$ and a the radius of the sphere. It is obvious that $R_t(\eta)$ does not depend on time whenever the particle undergoes a purely translational motion, while, for more complicated motions, it is responsible for the amplitude variations of the diffused field.

For some significant shapes (ellipsoids, cylinders and so on) the factor R_t is given by means of simple analytical expressions in terms of suitable parameters that define the orientation of the scatterer with respect to the incident field (see van de Hulst, 1957, Chapter VIII). Thus the dependence on time of these parameters as a function of the scatterer's motion determines $R_t(\eta)$ explicitly.

The intensity of light scattered in a given direction η is easily evaluated by means of Eqs. (6-7) and (6-10) as

$$I_1(\eta) = \frac{c}{8\pi} |\hat{E}_1(t)|^2 = \frac{I_0 k_0^4 \sin^2 \gamma}{(2\pi r)^2} n_0^2 (n - n_0)^2 |R_t(\eta)|^2 \tag{6-13}$$

For unpolarized (natural) incident light, one has to average over the possible directions of $\mathbf{E}_0$ which results in the substitution $\sin^2 \gamma \rightarrow \frac{1}{2}(1 + \cos^2 \vartheta)$, where γ and ϑ are the angles between $\mathbf{E}_0$ and $\mathbf{k}_1$ and $\mathbf{k}_0$ and $\mathbf{k}_1$ (see Fig. 6.2). The expression obtained in this way can be compared with the analogous one worked out in the static case (van de Hulst, 1957, Chapter VII).

As already mentioned, the ratio Q_{sc} between the total electromagnetic energy scattered per unit time, which can be evaluated

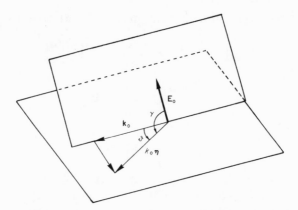

Fig. 6.2   *Relevant angles in the scattering geometry.*

by integrating Eq. (6-13) over a spherical surface of radius r and the energy flux incident on the particle turns out to be much less than unity (van de Hulst, 1957, Chapter VII), which justifies the macroscopic single scattering approach on which Eq. (6-4) is based. We observe in this framework that single scattering means that the incident radiation is scattered only once by the isolated particle, a condition different from the analogous one pertaining to an ensemble of particles. Whenever the Rayleigh–Gans criterion, that is, the single-scattering condition for the single scatterer, is not satisfied, no simple general theory exists. One can, however, still write in full generality

$$\hat{\mathbf{E}}_1(t) = \mathbf{A}_i(t)\exp\{i[\mathbf{k}_1 \cdot \mathbf{r}_i(t) - \omega_0 t]\} \qquad (6\text{-}14)$$

where $\mathbf{r}_i(t)$ is the trajectory of a given point of the particle and $\mathbf{A}_i(t)$ (which is in general the product of a real vector times a phase factor) depends on the rotational motion around $\mathbf{r}_i(t)$ and is independent of time if such a motion is not present. In particular, $\mathbf{A}_i(t)$ is time independent for a sphere, if $\mathbf{r}_i(t)$ represents the motion of its center. The evaluation of $\mathbf{A}_i(t)$ is in this case accomplished in the framework of the so-called *Mie theory* of scattering (see, for example, Born and Wolf, 1970,

Chapter XIII). Whenever the particles are small compared with the incident wavelength (Tyndall scattering), we have

$$\mathbf{A}_i(t) = -\frac{k_0^2 \, \boldsymbol{\eta} \times [\boldsymbol{\eta} \times \alpha(t) \cdot \mathbf{E}_0]}{r} \qquad (6\text{-}15)$$

where the polarizability tensor $\alpha(t)$ has been evaluated for some simple situations [spheres, ellipsoids, and some others (see van de Hulst, 1957, Chapter VI)].

In particular, for a sphere of radius a, α reduces to a scalar given by

$$\alpha = \frac{n^2 - n_0^2}{n^2 + 2n_0^2} a^3 \qquad (6\text{-}16)$$

We can summarize the results of this section by considering two classes of scattering particles, according to whether or not the Rayleigh–Gans criterion is satisfied. In the first case, the conditions for single scattering are met. This in turn implies that the direction of $\hat{\mathbf{E}}_1(\mathbf{r}, t)$ does not depend on the orientation of the scatterer as implied by the constancy of $\mathbf{B}$ in Eq. (6-8). The amplitude depends on time, unless the dimensions of the scatterer are small compared with the wavelength. When the Rayleigh–Gans criterion is not satisfied, both direction and amplitude of $\hat{\mathbf{E}}_1(\mathbf{r}, t)$, that is, of $\mathbf{A}_i(t)$ in Eq. (6-14), depend on the particle orientation, irrespective of its dimensions. Spherical particles, for which the scattered field is always constant in amplitude and direction, are in a class by themselves.

VI.3 Scattering by an Ensemble of Particles

The usual situations are those in which light scattering takes place in a volume containing a large number of particles. We shall next consider the case of an ensemble of N identical particles and neglect the fluctuations of N in the scattering volume. We shall treat separately situations in which these fluctuations play a fundamental role (see Section VI.8).

The position and orientation of the ith scatterer are described by $\mathbf{r}_i(t)$ and the three orientational Eulerian angles $\Omega_i(t) \equiv (\Omega_{1i}, \Omega_{2i}, \Omega_{3i})$, so that the electric field singly scattered[†] by the ensemble can be written as

$$\hat{\mathbf{E}}_i(\mathbf{r}, t) = \sum_{i=1}^{N} \mathbf{A}[\Omega_i(t)] \exp\{i[\mathbf{k}_i \cdot \mathbf{r}_i(t) - \omega_0 t]\} \quad (6\text{-}17)$$

We can use Eq. (6-17) in order to determine the spectrum $I(\omega)$ of the scattered field [see Eq. (2-43)]. We must evaluate, in a stationary situation, the ensemble average

$$\langle \hat{\mathbf{E}}(t+\tau) \cdot \hat{\mathbf{E}}^*(t) \rangle = \sum_{i=1}^{N} \sum_{j=1}^{N} \langle \mathbf{A}[\Omega_i(t+\tau)] \cdot \mathbf{A}^*[\Omega_j(t)]$$
$$\times \exp\{i\mathbf{k}_1 \cdot [\mathbf{r}_i(t+\tau) - \mathbf{r}_j(t)]\} \rangle e^{-i\omega_0 \tau} \quad (6\text{-}18)$$

where the angle brackets stand for statistical averages over the translational and rotational motion of the particles. Whenever the two motions are decoupled (this is not always the case; see the end of Section VI.5) we can rewrite

$$\langle \hat{\mathbf{E}}(t+\tau) \cdot \hat{\mathbf{E}}^*(t) \rangle = \sum_{i=1}^{N} \sum_{j=1}^{N} \langle \mathbf{A}[\Omega_i(t+\tau)] \cdot \mathbf{A}^*[\Omega_j(t)] \rangle$$
$$\times \langle \exp\{i\mathbf{k}_1 \cdot [\mathbf{r}_i(t+\tau) - \mathbf{r}_j(t)]\} \rangle e^{-i\omega_0 \tau} \quad (6\text{-}19)$$

It is convenient at this point to underline the particular meaning of the ensemble-averaging operation within the framework of the developments of this chapter. In fact, there is a

---

[†] We observe that *single-scattered field* means here that each particle *sees* the unperturbed incident field. This in turn implies that the ratio between the total electromagnetic intensity scattered per unit time by all particles and the energy flux incident on the scattering volume is small compared to unity, that is, $NQ^{sc}\sigma/S \simeq \sigma\rho d \ll 1$ where ρ is the density of scatterers, σ and S being the typical sections of the scatterer and of the scattering volume of length d, respectively.

fundamental difference between the present situation and the one pertinent to molecular scattering previously treated in which the forces acting on each scatterer are only the ones exerted by the other scatterers. In this case, each sample of the statistical ensemble over which the average operation has to be performed is completely determined by the initial position and velocity of each molecule. This information, however, is not sufficient to characterize each sample if one deals with particles embedded in a medium that undergoes random fluctuations. The dynamics of the scatterers are now significantly influenced by the medium itself, so that the possible realizations of the statistical ensemble must include those of the supporting medium. In other words, one must average over the initial positions and velocities of the particles and over the fluctuating quantities, relative to the medium, that influence the particle motion. Accordingly, we shall indicate in the following with the symbols $\langle\cdots\rangle$, $\langle\cdots\rangle_P$, and $\langle\cdots\rangle_M$ the operations of complete ensemble averaging, averaging on the particles, and averaging over the host medium. A far-reaching consequence of the presence of a supporting medium is, as we shall see, the possibility of a relevant statistical dependence among the particles, notwithstanding the absence of direct mutual interaction.

Equation (6-19) can be suitably simplified if the medium is homogeneous. In a stationary and homogeneous situation, which implies a stationary medium and a uniform probability distribution for the initial positions of the particles, we can write

$$\langle \hat{\mathbf{E}}(t+\tau)\cdot\hat{\mathbf{E}}^*(t)\rangle = \langle\hat{\mathbf{E}}(\tau)\cdot\hat{\mathbf{E}}^*(0)\rangle$$

$$= \sum_{i=1}^{N}\sum_{j=1}^{N}\langle \mathbf{A}[\Omega_i(\tau)]\cdot\mathbf{A}^*[\Omega_j(0)]\rangle$$

$$\times\langle\exp\{i\mathbf{k}_1\cdot[\mathbf{r}_i(\tau)-\mathbf{r}_j(0)]\}\rangle e^{-i\omega_0\tau} \qquad (6\text{-}20)$$

Let us now consider the generic term $\langle\exp\{i\mathbf{k}_1\cdot[\mathbf{r}_i(\tau)-\mathbf{r}_j(0)]\}\rangle$

and recall the obvious relation

$$\mathbf{r}_i(t) = \mathbf{r}_{i0} + \int_0^t \mathbf{v}_i(t'; \mathbf{r}_{i0}, \mathbf{v}_{i0})\, dt' \tag{6-21}$$

where

$$\mathbf{v}_i(t) = \dot{\mathbf{r}}_i(t) \tag{6-22}$$

and

$$\mathbf{r}_{i0} = \mathbf{r}_i(t = 0), \qquad \mathbf{v}_{i0} = \mathbf{v}_i(t = 0) \tag{6-23}$$

We then have

$$\langle \exp\{i\mathbf{k}_1 \cdot [\mathbf{r}_i(\tau) - \mathbf{r}_j(0)]\}\rangle$$

$$= \left\langle \exp[i\mathbf{k}_1 \cdot (\mathbf{r}_{i0} - \mathbf{r}_{j0})] \exp\left[i\mathbf{k}_1 \cdot \int_0^\tau \mathbf{v}_i(t'; \mathbf{r}_{i0}, \mathbf{v}_{i0})\, dt' \right] \right\rangle \tag{6-24}$$

The ensemble average over the medium is, due to homogeneity, independent of $\mathbf{r}_{i0}$ so there remains to average over the initial distribution of the positions of the ith and jth particles only the function $\exp[i\mathbf{k}_1 \cdot (\mathbf{r}_{i0} - \mathbf{r}_{j0})]$. The uniformity of the probability distribution ensures then that the right-hand side of Eq. (6-24) vanishes for $i \neq j$. Since for $i = j$, the right-hand side of Eq. (6-24) is independent of the particular scatterer, Eq. (6-20) reduces to

$$\langle \hat{\mathbf{E}}(\tau) \cdot \hat{\mathbf{E}}^*(0)\rangle = N\langle \mathbf{A}[\Omega_i(\tau)] \cdot \mathbf{A}^*[\Omega_i(0)]\rangle$$

$$\times \langle \exp\{i\mathbf{k}_1 \cdot [\mathbf{r}_i(\tau) - \mathbf{r}_{i0}]\}\rangle_M \, e^{-i\omega_0 t} \tag{6-25}$$

where the subscript i represents the generic particle. Thus we obtain from Eqs. (2-48) and (6-25) the expression for the spectrum of the scattered field in the form of the convolution integral

$$I(\omega) = N\frac{c}{8\pi} \int_{-\infty}^{+\infty} d\omega'\, \tilde{C}_R(\omega')\, \tilde{C}_T(\omega - \omega') \tag{6-26}$$

where $\tilde{C}_R(\omega)$ and $\tilde{C}_T(\omega)$ are the Fourier transforms of the rotational and translational autocorrelation functions

$$C_R(\tau) = \langle \mathbf{A}[\Omega_i(\tau)] \cdot \mathbf{A}^*[\Omega_i(0)] \rangle \tag{6-27}$$

and

$$C_T(\tau) = \langle \exp\{i\mathbf{k}_1 \cdot [\mathbf{r}_i(\tau) - \mathbf{r}_{i0}]\} \rangle_M \tag{6-28}$$

Although the combined effects of both translational and rotational motions on the form of the spectrum can be relevant (see the end of Section VI.5), we shall mainly limit ourselves in the following to consideration of the cases of spherical particles of arbitrary size and of Rayleigh–Gans particles, whose size is much‑smaller than the incident wavelength. Then, $\mathbf{A}[\Omega_i(t)]$ is time independent and the rotational dynamics have no influence on the scattered radiation. We note that, under this assumption, the scattered field reduces to

$$\hat{\mathbf{E}}_1(r,t) = \mathbf{A}(\mathbf{k}_1) \sum_{i=1}^{N} \exp\{i[\mathbf{k}_1 \cdot \mathbf{r}_i(t) - \omega_0 t]\}$$

$$= \mathbf{A}(\mathbf{k}_1)\tilde{n}_1(-\mathbf{k}_1, t) \tag{6-29}$$

where $\tilde{n}_1(-\mathbf{k}_1, t)$ is the Fourier space transform of

$$n_1(\mathbf{r}, t) = \sum_{i=1}^{N} \delta[\mathbf{r} - \mathbf{r}_i(t)] \tag{6-30}$$

which can be interpreted as the microscopic density of the geometrical points representing the translational particle motion.

The preceding considerations refer to the first-order statistical properties of the scattered radiation. We are now in a position to apply them to specific situations and to extend them, case by case, to include higher-order statistical properties.

VI.4 Scattering by Particles Suspended in a Laminar Flow

A basic physical situation, particularly important for the present day widely developed technique of *laser anemometry* to which it gave rise, is the one relative to a fluid possessing a laminar motion. We assume hereafter that the suspended particles follow exactly the motion of the fluid elements surrounding them, a hypothesis that can be justified when the particles are very small with respect to the smallest scale of the fluid velocity variation (see Hinze, 1959, Chapter V). It is similarly assumed that the particles do not influence the motion of the fluid. Under the usual condition of stationarity for the velocity field we have [see Eq. (6-29)]

$$\langle \hat{\mathbf{E}}(t+\tau) \cdot \hat{\mathbf{E}}^*(t) \rangle = \langle \hat{\mathbf{E}}(\tau) \cdot \hat{\mathbf{E}}^*(0) \rangle$$

$$= |\mathbf{A}|^2 \sum_{i=1}^{N} \sum_{j=1}^{N} \langle \exp[i\mathbf{k}_1 \cdot \mathbf{r}_i(\tau) - i\mathbf{k}_1 \cdot \mathbf{r}_{j0}] \rangle$$
$$\times e^{-i\omega_0 \tau} \tag{6-31}$$

If the velocity field is indicated by $\mathbf{U}(\mathbf{r})$, we have

$$\mathbf{r}_i(t) = \mathbf{r}(t; \mathbf{r}_{i0}) = \mathbf{r}_{i0} + \int_0^t \mathbf{U}[\mathbf{r}(t'; \mathbf{r}_{i0})] \, dt' \tag{6-32}$$

so that

$$\langle \hat{\mathbf{E}}(\tau) \cdot \hat{\mathbf{E}}^*(0) \rangle = |\mathbf{A}|^2 \sum_{i,j} \langle \exp[i\mathbf{k}_1 \cdot (\mathbf{r}_{i0} - \mathbf{r}_{j0})]$$
$$\times \exp\{i\mathbf{k}_1 \cdot \int_0^\tau \mathbf{U}[\mathbf{r}(t'; \mathbf{r}_{i0})] \, dt'\} \rangle e^{-i\omega_0 \tau} \tag{6-33}$$

The ensemble-averaging operation must be performed only on the initial-particle positions due to the deterministic nature of the motion of the fluid. We must note that the velocity field is in general not homogeneous, a fact that is not always consistent with the assumption of a uniform distribution of the initial positions of the particles. In any case, if the initial probability distribution can be assumed to vary over a scale much larger than $1/k_1$, the terms with $i \neq j$ vanish in Eq. (6-33) so that

$$\langle \hat{\mathbf{E}}(\tau) \cdot \hat{\mathbf{E}}^*(0) \rangle = |\mathbf{A}|^2 N \langle \exp\{i\mathbf{k}_1 \cdot \int_0^\tau \mathbf{U}[\mathbf{r}(t'; \mathbf{r}_{i0})]\, dt'\} \rangle_P\, e^{-i\omega_0\tau}$$

(6-34)

Let us consider for definiteness the practical situation in which a liquid is made to flow in a tube of constant section so that the radial velocity profile obeys the usual parabolic law (see Fig. 6.3)

$$U_z(\rho) = C(R^2 - \rho^2) \tag{6-35}$$

where $\mathbf{r} = (\rho, z)$, R the radius of the tube, and C a constant. Equation (6-34) reduces to

$$\langle \hat{\mathbf{E}}(\tau) \cdot \hat{\mathbf{E}}^*(0) \rangle = |\mathbf{A}|^2 N \langle \exp[ik_{1z} U_z(\rho) \tau] \rangle_P\, e^{-i\omega_0\tau}$$

$$= |\mathbf{A}|^2 N e^{-i\omega_0\tau} \frac{2\pi}{V_{sc}} \int_{V_{sc}} \exp[ik_{1z} U_z(\rho)]\, \rho\, d\rho\, dz$$

(6-36)

the evaluation of the integral requiring the knowledge of the shape of the scattering volume V_{sc}. If the transverse dimension of V_{sc} is small compared with the tube diameter, Eq. (6-36) reduces to

$$\langle \hat{\mathbf{E}}(\tau) \cdot \hat{\mathbf{E}}^*(0) \rangle = |\mathbf{A}|^2 N \exp\{i[k_{1z} U_z(\bar{\rho}) - \omega_0] \tau\} \quad \text{(6-37)}$$

where $\bar{\rho}$ indicates a typical transverse coordinate of V_{sc}.

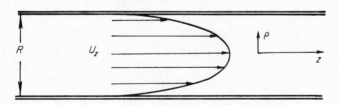

Fig. 6.3   *Parabolic velocity profile.*

In order to gain some insight into the higher-order statistical properties of the scattered field, let us evaluate the second-order correlation function of the field

$$\langle |\hat{\mathbf{E}}(\tau)|^2 \, |\hat{\mathbf{E}}(0)|^2 \rangle$$

$$
\begin{aligned}
= |\mathbf{A}|^4 \sum_{i,j,l,m} \Bigg\langle & \exp\{i\mathbf{k}_1 \cdot (\mathbf{r}_{i0} - \mathbf{r}_{j0} + \mathbf{r}_{l0} - \mathbf{r}_{m0})\} \\
& \times \exp\left\{ i\mathbf{k}_1 \cdot \int_0^\tau \mathbf{U}[\mathbf{r}(t';\mathbf{r}_{i0})]\, dt' \right\} \\
& \times \exp\left\{ -i\mathbf{k}_1 \cdot \int_0^\tau \mathbf{U}[\mathbf{r}(t';\mathbf{r}_{j0})]\, dt' \right\} \Bigg\rangle_{\!\!P}
\end{aligned}
\tag{6-38}
$$

Equation (6-38) can be written as

$$\langle |\hat{\mathbf{E}}(\tau)|^2 \, |\hat{\mathbf{E}}(0)|^2 \rangle$$

$$
\begin{aligned}
= |\mathbf{A}|^4 \sum_{i,j} \Bigg\langle & \exp[i\mathbf{k}_1 \cdot (\mathbf{r}_{i0} - \mathbf{r}_{j0})] \exp\left\{ i\mathbf{k}_1 \cdot \int_0^\tau \mathbf{U}[\mathbf{r}(t';\mathbf{r}_{i0})]\, dt' \right\} \\
& \times \exp\left\{ -i\mathbf{k}_1 \cdot \int_0^\tau \mathbf{U}[\mathbf{r}(t';\mathbf{r}_{j0})]\, dt' \right\} \Bigg\rangle_{\!\!P} \\
+ |\mathbf{A}|^4 \sum_{i \neq j} \Bigg\langle & \exp\left\{ i\mathbf{k}_1 \cdot \int_0^\tau \mathbf{U}[\mathbf{r}(t';\mathbf{r}_{i0})]\, dt' \right\} \\
& \times \exp\left\{ -i\mathbf{k}_1 \cdot \int_0^\tau \mathbf{U}[\mathbf{r}(t';\mathbf{r}_{j0})]\, dt' \right\} \Bigg\rangle_{\!\!P}
\end{aligned}
\tag{6-39}
$$

where terms containing factors of the type $\langle \exp(i\mathbf{k}_1 \cdot \mathbf{r}_{l0}) \rangle_P$ and $\langle \exp(-i\mathbf{k}_1 \cdot \mathbf{r}_{m0}) \rangle_P$ have been neglected due to the uniformity of the initial probability distribution over a scale much larger than $1/k_1$. The same argument can be used to ignore in the first term on the right-hand side of Eq. (6-39) the contribution with $i \neq j$, if the velocity field $\mathbf{U}(\mathbf{r})$ behaves in such a way that the quantity $\langle \exp\{i\mathbf{k}_1 \cdot \int_0^\tau \mathbf{U}[\mathbf{r}(t',\mathbf{r}_{i0})]\, dt' \rangle_P$ does not significantly vary as a function of $\mathbf{r}_{i0}$ over a distance $1/k_1$. Thus with the help of Eq. (6-36), Eq. (6-39) reduces to

$$\langle |\hat{E}(\tau)|^2 |\hat{E}(0)|^2 \rangle = \langle |\hat{E}(0)|^2 \rangle_P^2 + |\langle \hat{E}(\tau) \cdot \hat{E}^*(0) \rangle_P|^2 \quad (6\text{-}40)$$

having disregarded contributions of the order of $1/N$.

The factorization property of the second-order correlation function shown by Eq. (6-40) is characteristic of a Gaussian field [see Eq. (4-110)] and can be easily generalized to higher orders. This circumstance, which on the other hand was to be expected a priori on the basis of the central limit theorem (see Section V.4) due to the complete mutual statistical independence of the scatterers, ensures that all information is exhausted by first-order measurements. Thus, for example, Eq. (6-37) provides for the power spectrum

$$I(\omega) = \frac{c}{8\pi} N |A|^2 \delta [\omega - \{\omega_0 - k_{1z} U_z(\bar{\rho})\}] \quad (6\text{-}41)$$

which describes the Doppler shift relative to the particles moving with velocity $U_z(\bar{\rho})$. In this case a second-order measurement does not contain any information on the magnitude of the velocity, since $|\langle \hat{E}(\tau) \cdot \hat{E}^*(0) \rangle_P|^2 = |\langle \hat{E}(0) \cdot \hat{E}^*(0) \rangle_P|^2$, and could be regarded only as a test of the uniformity of the velocity field over the scattering volume.

An experimental investigation of the Doppler shift of scattered light by particles suspended in a laminar flow was first carried out in a liquid by Yeh and Cummins (1964) and successively repeated in air by Foreman *et al.* (1965, 1966). Some experimental results obtained by means of the heterodyne technique are shown in Figs. 6.4 and 6.5.

Since then, this technique has undergone extensive development to become now-standard *laser anemometry* (see, for example, Durst *et al.*, 1972; Abbiss *et al.*, 1974).

We wish finally to observe that in the preceding considerations the possible contribution to the spectrum of the scattered light due to the finite transit time $\bar{t}$ of the particle in the scattering volume has been neglected. This is justified whenever the frequency range of the scattered field one has to observe is

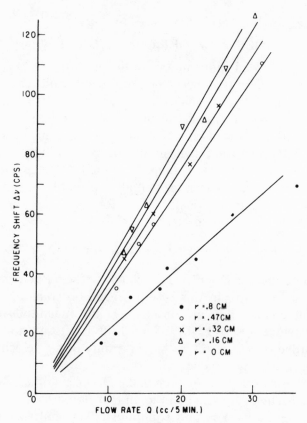

Fig. 6.4 *Experimental frequency shifts at several radial positions as a function of flow rate (after Yeh and Cummins, 1964).*

much larger than $1/\bar{t}$ (for a detailed discussion see Pike *et al.*, 1968).

VI.5 Scattering by Particles Undergoing Brownian Motion

In the previous section, we completely ignored the fact that the particles undergo, besides the translational motion associated

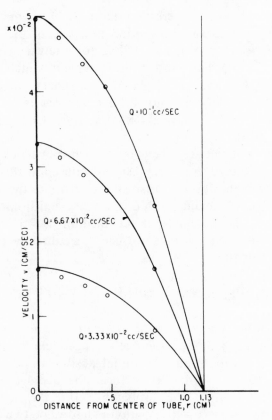

Fig. 6.5   *Velocity profiles for three flow rates (after Yeh and Cummins, 1964). The solid curves are theoretical; the points are experimental.*

with the flow of the fluid, a diffusive motion due to collisions with molecules of the surrounding medium. This phenomenon, *Brownian motion*, gives rise to a frequency broadening of the scattered field, which in turn establishes minimum values of detectable velocity and velocity difference in the laser anemometry technique. On the other hand, the observation of Brownian motion by study of its influence on the spectrum of the scattered light is interesting in itself as a method for gaining information on the motion and size of the particles and on the viscosity of the host fluid (or equivalently on the diffusion constant). This kind of measurement, which was first performed by Cummins *et al.* (1964) for ordinary particles and by Dubin *et al.* (1967) for biological molecules, is, at present, widely used (see, for example, Benedek, 1969; Chu, 1970; and Cummins, 1974).

We shall treat the scattering by Brownian particles in the framework of the general formalism introduced in this chapter including the higher-order statistical properties of the scattered field. Since we are dealing with a stationary and homogeneous situation and the influence of the medium on the particles possesses a stochastic nature, we can start directly from Eq. (6-34), conveniently rewritten as

$$\langle \hat{\mathbf{E}}(\tau) \cdot \hat{\mathbf{E}}^*(0) \rangle = N|\mathbf{A}|^2 \langle \exp\{i\mathbf{k}_1 \cdot [\mathbf{r}(\tau) - \mathbf{r}_0]\} \rangle e^{i\omega_0 \tau} \quad (6\text{-}42)$$

where $\mathbf{r}(\tau)$ and $\mathbf{r}_0$ are the position at time $t = \tau$ and $t = 0$ of the generical particle, and the ensemble average includes averaging over both the medium and the initial positions and velocities of the particles. In order to perform the average over the medium we must use the probability $W(\mathbf{r}, \tau; \mathbf{r}_0, \mathbf{u}_0)$ that a particle started at $t = 0$ with position $\mathbf{r}_0$ and velocity $\mathbf{u}_0$ reaches the position $\mathbf{r}$, at time τ, thus obtaining

$$\langle \exp\{i\mathbf{k}_1 \cdot [\mathbf{r}(\tau) - \mathbf{r}_0]\} \rangle_{\mathrm{M}} = \int \exp[i\mathbf{k}_1 \cdot (\mathbf{r} - \mathbf{r}_0)] \, W(\mathbf{r}, \tau; \mathbf{r}_0, \mathbf{u}_0) \, d\mathbf{r}_0$$

$$(6\text{-}43)$$

The resulting expression must be averaged over the initial position and velocity probability distribution of the particle, the first average being immaterial since W depends on space only through $\mathbf{r} - \mathbf{r}_0$ so that the right-hand side of Eq. (6-43) does not depend on $\mathbf{r}_0$. If $P(\mathbf{u}_0)$ indicates the initial velocity probability distribution, we have

$$\langle \exp\{i\mathbf{k}_1 \cdot [\mathbf{r}(\tau) - \mathbf{r}_0]\} \rangle = \int \exp[i\mathbf{k}_1 \cdot (\mathbf{r} - \mathbf{r}_0)]\, W(\mathbf{r}, \tau; \mathbf{r}_0, \mathbf{u}_0)$$

$$\times\, P(\mathbf{u}_0)\, d\mathbf{r}_0\, d\mathbf{u}_0 \qquad (6\text{-}44)$$

The explicit forms of W and P can be found, for example, in the classical paper of Chandrasekhar (1943). They are both Gaussian, $P(\mathbf{u}_0)$ corresponding, in particular to the Maxwellian velocity distribution apt to describe our equilibrium situation. Thus performing the integration over $\mathbf{u}_0$ in Eq. (6-44), we obtain

$$\langle \exp\{i\mathbf{k}_1 \cdot [\mathbf{r}(\tau) - \mathbf{r}_0]\} \rangle = \int e^{i\mathbf{k}_1 \cdot \boldsymbol{\rho}}\, G(\boldsymbol{\rho})\, d\boldsymbol{\rho} \qquad (6\text{-}45)$$

where $G(\boldsymbol{\rho})$ is a Gaussian probability distribution with zero average and variance

$$\langle \rho^2 \rangle = \frac{6K_\mathrm{B} T}{m\beta^2}(\beta\tau - 1 + e^{-\beta\tau}) \qquad (6\text{-}46)$$

[see Eq. (125′) of Chandrasekhar (1943)]. Here K_B is Boltzmann's constant, m and a the mass and radius of the particle, and $\beta = 6\pi a\eta/m$, η representing the coefficient of viscosity at temperature T. For such a distribution it holds that

$$\langle e^{i\mathbf{k}_1 \cdot \boldsymbol{\rho}} \rangle = \langle e^{ik_{1x}x} \rangle \langle e^{ik_{1y}y} \rangle \langle e^{ik_{1z}z} \rangle$$

$$= \exp\left(-\frac{1}{6}k_1^2 \langle \rho^2 \rangle\right) = \exp\left[-k_1^2 \frac{K_\mathrm{B} T}{m\beta^2}(\beta\tau - 1 + e^{-\beta\tau})\right]$$

$$(6\text{-}47)$$

having taken into account that for a normally distributed variable Ψ we have

$$\langle e^{i\Psi} \rangle = e^{-\langle \Psi^2 \rangle/2} \qquad (6\text{-}48)$$

By using Eqs. (6-42) and (6-47) we can then evaluate the optical spectrum of the scattered light as

$$
I(\omega) = \frac{c}{16\pi^2} \int_{-\infty}^{+\infty} \langle \hat{\mathbf{E}}(\tau) \cdot \hat{\mathbf{E}}^*(0) \rangle \, e^{i\omega\tau} \, d\tau
$$

$$
= \frac{N|\mathbf{A}|^2 c}{16\pi^2} \int_{-\infty}^{+\infty} \exp\left[-k_1^2 \frac{K_B T}{m\beta^2}(\beta\tau - 1 + e^{-\beta\tau}) \right] e^{i(\omega - \omega_0)\tau} \, d\tau
$$

$$(6\text{-}49)$$

Since $k_1^2(K_B T/m\beta^2)$ is in practice much smaller than unity, we can correctly ignore the quantity $e^{-\beta\tau} - 1$ in the exponential, thus obtaining the Lorentzian spectrum

$$
I(\omega) = \frac{cN|\mathbf{A}|^2}{16\pi^2} \frac{k_1^2 D}{(\omega - \omega_0)^2 + (k_1^2 D)^2} \tag{6-50}
$$

with

$$
D = \frac{K_B T}{m\beta}
$$

This expression was obtained first by Glauber (1962) and Pecora (1964), and by many authors since (see, for example, Cummins *et al.*, 1964; Arecchi *et al.*, 1967; Clark *et al.*, 1970).

As in the previous section, the Gaussian factorization of the higher-order correlation functions of the scattered field is a consequence of the central limit theorem. In particular,

$$
\langle |\hat{\mathbf{E}}(\tau)|^2 |\hat{\mathbf{E}}(0)|^2 \rangle = \langle |\hat{\mathbf{E}}(0)|^2 \rangle^2 + |\langle \hat{\mathbf{E}}(\tau) \cdot \hat{\mathbf{E}}^*(0) \rangle|^2 \tag{6-51}
$$

so that a second-order measurement contains the same information as a first-order one due to the reality of the quantity $\langle \hat{\mathbf{E}}(\tau) \cdot \hat{\mathbf{E}}^*(0) \rangle$ [apart from the irrelevant factor $\exp(-i\omega_0 \tau)$]. As a matter of fact, the method of self-beating spectroscopy is successfully applied in measuring the diffusion coefficient D of the particles (Dubin *et al.*, 1967).

An experimental verification of the Gaussian nature of the scattered light requires measurements of both $\langle |\hat{\mathbf{E}}(\tau)|^2 |\hat{\mathbf{E}}(0)|^2 \rangle$ and $\langle \hat{\mathbf{E}}(\tau) \cdot \hat{\mathbf{E}}^*(0) \rangle$. In particular, the frequency dependence of the intensity-fluctuation spectrum $P(\omega)$ must be, according to Eq. (5-71), Lorentzian, centered around zero frequency with half-width $2\Delta\omega$ two times larger than $\Delta\omega = k_1^2 D$ relevant to the power spectrum. Thus a necessary condition for the validity of Eq. (6-51) is verified by measuring, by means of heterodyne and self-beating techniques, the spectral profiles of scattered light and comparing the relative half-widths. Figure 6.6 shows the results obtained by Cummins *et al.* (1969) on a solution of polystyrene latex spheres displaying both the correct relation between the two half-widths and their angular dependence on k_1^2.

The direct proof of the Gaussian behavior of the scattered

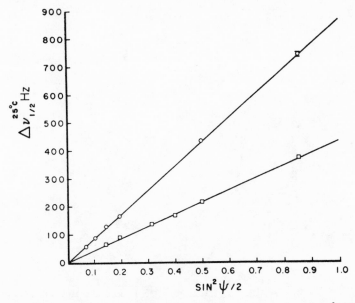

Fig. 6.6 *Single Lorentzian half-widths at 25° C as a function of* $\sin^2(\psi/2)$ *(ψ here is the scattering angle) obtained from self-beating (○) and heterodyne (□) spectra of polystyrene latex spheres (0.126 ± 0.004 µm diameter) (after Cummins et al., 1969).*

field has been obtained by measuring the photocount distribution $p(n, T)$ [see Eq. (4-105)] at a given angle. More precisely, Arecchi *et al.* (1967) have tested the expected Bose–Einstein distribution [see Eq. (4-111)] valid for counting times such that $(\Delta\omega/2\pi)T = \Gamma T \ll 1$ (see Fig. 6-7), while Jakeman *et al.* (1968) have performed measurements up to counting times T such that $\Gamma T \simeq 1.6$. In Fig. 6.8 a typical photon-counting distribution is reported. In Fig. 6.9 the normalized factorial moments

$$\frac{M_k}{\langle m \rangle^k} = \frac{m(m-1)\cdots(m-k+1)}{\langle m \rangle^k}$$

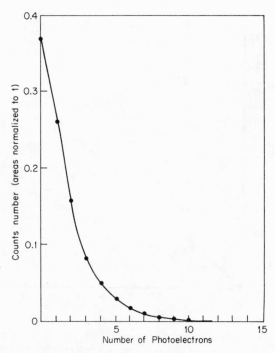

Fig. 6.7   *Typical photon-counting distribution of light scattered by polystyrene spherical particles (1.0088 μm diameter) (after Arecchi et al., 1967).*

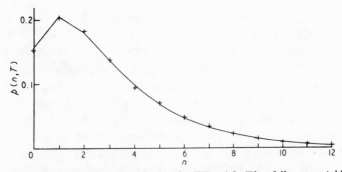

Fig. 6.8 *Photon-counting distribution for $\Gamma T = 1.0$. The full curve yields the theoretical behavior and the crosses are the experimental results (after Jakeman et al., 1968).*

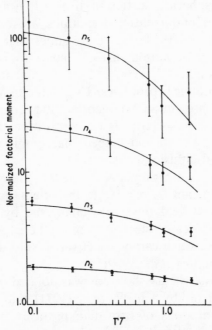

Fig. 6.9 *Normalized factorial moments of photon-counting distribution obtained with six different values of sample time T and a fixed linewidth of 20 Hz. The curves yield the theoretical behavior and the filled circles are the experimental results (after Jakeman et al., 1968).*

are displayed as a function of the counting time. We observe that in the limit of small ΓT the normalized kth factorial moment approaches $k!$ according to Eq. (4-112), while for large ΓT, it tends to unity. This last fact can be easily understood recalling that any stochastic stationary field possesses a Poisson photocount distribution for times much larger than the coherence time, since in this limit the fluctuations average to zero so that the field behaves as a well-prescribed field [see Eqs. (4-106) and (4-106)′]. These considerations are in agreement with the results of Figs. 6.8 and 6.9 which describe a behavior intermediate between Bose–Einstein and Poisson.

Rotational Brownian motion can affect the scattered field in the case of nonspherical scatterers. If as a first approximation the translational and rotational motions are considered decoupled (see Pecora, 1964), Eq. (6-26) states that the scattered spectrum is the convolution of the Fourier transforms of the correlation functions $C_T(\tau)$ and $C_R(\tau)$ pertinent to the two independent dynamical processes. Under the assumption that $C_R(\tau)$ and $C_T(\tau)$ present an exponential time decay with one or more decay times, the spectrum is a superposition of Lorentzians. As an example, a rigid rod-shaped macromolecule possesses two translational diffusion constants $D_{\parallel}$ and $D_{\perp}$ relative to the motions parallel and orthogonal to the axis of symmetry and one diffusion constant D_R relative to the orientation of the axis with the scattered field obviously unaffected by the rotation around it. In this case, the spectrum is the superposition of two Lorentzians, which can be effectively resolved under some conditions (Cummins *et al.*, 1969).

A more rigorous theory shows, as was pointed out by Maeda and Saito (1969) and by Schaefer *et al.* (1971), that translational and rotational motions of Brownian particles are in general coupled, the decoupling being legitimate for an axisymmetric particle moving parallel to its axis of symmetry (Chow, 1973). Experimental deviations from the spectral behavior suggested by the theory of Pecora (1964) have been actually found by Fujima (1969) and by Schaefer *et al.* (1971).

VI.6 Scattering by Particles Suspended in a Turbulent Fluid

The scattering of light by particles suspended in a fluid undergoing a turbulent motion seems to be a natural extension of scattering by a laminar flow, the main difference lying in the stochastic character of the involved velocities. As a matter of fact, the statistical uncertainty of the fluctuating velocities associated with that of the initial particle positions, already present in the laminar case, would suggest a priori a Gaussian statistical behavior for the scattered field.

In effect, the theoretical analysis shows a significant departure from the Gaussian one connected with the correlation function of the velocity field. This situation, which constitutes the first example of a non-Gaussian scattered field (Di Porto *et al.*, 1969), possesses also a practical relevance since the measurement of the second-order correlation function of the scattered radiation allows the determination of the vortex dimension. Through a first-order measurement, conversely, one is, in practice, only able to determine the strength of the turbulence field, that is, the mean square value of the velocity fluctuation.

Let us consider a homogeneous, stationary, turbulent field and let us apply Eq. (6-33) to this situation. The velocity $U(r, t)$, which now depends explicitly on time, of the fluid element with position r at time t can be written as the sum of an average term U_0, independent of r and t, plus a fluctuating contribution $U'(r, t)$:

$$U(r, t) = U_0 + U'(r, t) \qquad (6\text{-}52)$$

so that Eq. (6-33) becomes

$$\langle \hat{\mathbf{E}}(\tau) \cdot \hat{\mathbf{E}}^*(0) \rangle = |A|^2 \sum_{i,j} \left\langle \exp[i\mathbf{k}_1 \cdot (\mathbf{r}_{i0} - \mathbf{r}_{j0})] \right.$$

$$\left. \times \exp\left\{ i\mathbf{k}_1 \cdot \int_0^\tau \mathbf{U}'[\mathbf{r}(t'; \mathbf{r}_{i0}), t'] \, dt' \right\} \right\rangle$$

$$\times \exp\{i[\mathbf{k}_1 \cdot \mathbf{U}_0 - \omega_0]\tau\} \tag{6-53}$$

By first performing the average over the velocity field we have, because of the homogeneity of turbulence and of the uniform distribution of the initial particle positions,

$$\langle \hat{\mathbf{E}}(\tau) \cdot \hat{\mathbf{E}}^*(0) \rangle = |A|^2 \, N \exp\{i[\mathbf{k}_1 \cdot \mathbf{U}_0 - \omega_0]\tau\}$$

$$\times \left\langle \exp\left\{ i\mathbf{k}_1 \cdot \int_0^\tau \mathbf{U}'[\mathbf{r}(t'; \mathbf{r}_{i0}), t'] \, dt' \right\} \right\rangle_{\mathrm{M}} \tag{6-54}$$

where the index i refers to an arbitrary particle started with an arbitrary initial position $\mathbf{r}_{i0}$. The ensemble average on the right-hand side of Eq. (6-54) can be cast into an explicit form under a suitable hypothesis on the turbulent field. More precisely, this is the so-called *joint Gaussian distribution hypothesis*, usually made in the theory of turbulence (see, for example, Hinze, 1959, Chapter III) generalized to different time velocity fluctuations in the form

$$\langle U_1[\mathbf{r}_i(t_1), t_1] \, U_1[\mathbf{r}_i(t_2), t_2] \cdots U_1[\mathbf{r}_i(t_{2n}), t_{2n}] \rangle_{\mathrm{M}}$$

$$= \sum_\pi \langle U_1[\mathbf{r}_i(t_1), t_1] \, U_1[\mathbf{r}_i(t_2), t_2] \rangle_{\mathrm{M}}$$

$$\cdots \langle U_1[\mathbf{r}_i(t_{2n-1}), t_{2n-1}] \, U_1[\mathbf{r}_i(t_{2n}), t_{2n}] \rangle_{\mathrm{M}} \tag{6-55}$$

where the sum is taken over all permutations of indices $1, 2, \ldots, 2n$ and U_1 represents the component of $\mathbf{U}'$ along a given axis ($\mathbf{k}_1$ in our case). the corresponding odd-order products vanishing.

It is easy to show that Eq. (6-55) implies the displacement of the generic particle $\int_0^\tau U_1[\mathbf{r}(t', \mathbf{r}_{i0}), t']\, dt'$ to be a Gaussian variable, so that Eq. (6-54) can be rewritten, with the help of Eq. (6-48), in the form

$$\langle \hat{\mathbf{E}}(\tau) \cdot \hat{\mathbf{E}}^*(0) \rangle = |A|^2 N \exp[i(\mathbf{k}_1 \cdot \mathbf{U}_0 - \omega_0)\tau]$$

$$\times \exp\left\{ -\tfrac{1}{2}k_1^2 \int_0^\tau \int_0^\tau \langle U_1[\mathbf{r}_i(t'), t'] U_1[\mathbf{r}_i(t''), t''] \rangle_M \right.$$

$$\left. \times dt'\, dt'' \right\} \qquad (6\text{-}56)$$

The ensemble average appearing in Eq. (6-56) can be rewritten as

$$\langle U_1[\mathbf{r}_i(t'), t'] U_1[\mathbf{r}_i(t''), t''] \rangle_M = \overline{U_1^2}\, R_L(t'' - t') \qquad (6\text{-}57)$$

having introduced the *Lagrangian correlation function* $R_L(t'' - t')$ (see, for example, Hinze, 1959, Chapter I; Leslie, 1973) between the velocities of a well-defined fluid element evaluated at different times, a quantity depending on t'' and t' through the difference $t'' - t' = \tau'$ due to the stationarity and homogeneity hypotheses. In particular, since

$$\overline{U_1^2} \equiv \langle U_1[\mathbf{r}_i(t'), t'] U_1[\mathbf{r}_i(t'), t'] \rangle_M$$

we have $R_L(0) = 1$.

The optical spectrum of scattered light can now be obtained by Fourier transforming Eq. (6-56). To this end, we observe that the temporal behavior of the right-hand side of Eq. (6-56) can be given in the limiting cases of time τ much smaller or larger than the correlation time τ^*, after which $R_L(\tau^*) \simeq 0$.

We have, respectively,

$$\int_0^\tau \int_0^\tau R_L(t'' - t')\, dt'\, dt'' = 2\int_0^\tau (\tau - \tau')\, R_L(\tau')\, d\tau'$$

$$= \begin{cases} \tau^2 & \text{for} \quad \tau \ll \tau^* \\ 2J_L\,\tau & \text{for} \quad \tau \gg \tau^* \end{cases} \qquad (6\text{-}58)$$

where the quantity

$$J_L = \int_0^{+\infty} R_L(\tau')\, d\tau' \qquad (6\text{-}59)$$

gives a measure of the largest time during which, on the average, a particle persists in a motion in a given direction. Thus if, as often realized in practice, $\tau^* \gg (k_1^2\, \overline{U_1^2})^{-1/2}$, Eqs. (2-48), (6-56), (6-57), and (6-58) yield

$$I(\omega) = \frac{c}{16\pi^2} |\mathbf{A}|^2 N \int_{-\infty}^{+\infty} d\tau \exp\{i[\omega - (\omega_0 - \mathbf{k}_1 \cdot \mathbf{U}_0)]\,\tau\}$$

$$\times \exp\left\{ -(\tfrac{1}{2}) k_1^2\, \overline{U_1^2} \int_0^\tau dt' \int_0^\tau dt''\, R_L(t'' - t') \right\}$$

$$= \frac{c}{16\pi^2} |\mathbf{A}|^2 N \int_{-\infty}^{+\infty} d\tau \exp\{i[\omega - (\omega_0 - \mathbf{k}_1 \cdot \mathbf{U}_0)]\,\tau\}$$

$$\times \exp(-\tfrac{1}{2} k_1^2\, \overline{U_1^2}\, \tau^2)$$

$$= \frac{c|\mathbf{A}|^2}{16\pi^2} N \left(\frac{2\pi}{k_1^2\, \overline{U_1^2}} \right)^{1/2} \exp - \frac{(\omega - \Omega_0)^2}{2k_1^2\, \overline{U_1^2}} \qquad (6\text{-}60)$$

where

$$\Omega_0 = \omega_0 - \mathbf{k}_1 \cdot \mathbf{U}_0 \qquad (6\text{-}61)$$

is the incident frequency Doppler shifted by an amount corresponding to the average velocity of the fluid. We emphasize that the Gaussian form of the spectrum, as given in Eq. (6-60) is a consequence not only of the normal distribution of the displacement of each particle but also of the condition $\tau^* \gg (k_1^2 \overline{U_1^2})^{-1/2}$. In this respect we observe that, in the case of Brownian motion, the displacement of the generical particle is also normally distributed, but the spectrum possesses a Lorentzian behavior due to the negligible correlation time $(\tau^* \simeq 0)$.

The first measurements on the spectrum of light scattered by particles suspended in a turbulent fluid were performed by Goldstein and Hagen (1967) and Pike *et al.* (1968) by means of the heterodyne technique. Their results are consistent with the Gaussian spectrum of Eq. (6-60) (see Fig. 6.10).

The preceding analysis shows that the spectrum of the scattered light supplies the value of $\overline{U_1^2}$ but does not depend on the spatial structure of the turbulent field. Conversely, the following analysis makes clear how this structure influences a second-order measurement.

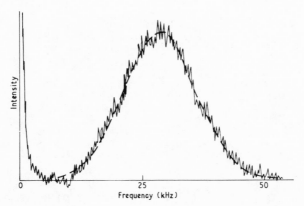

Fig. 6.10   *Heterodyne spectrum of light scattered by particles suspended in a turbulent flow. The profile has a Gaussian behavior (after Bourke et al., 1970).*

To this end, let us evaluate the second-order correlation function $\langle |\hat{\mathbf{E}}(\tau)|^2 \, |\hat{\mathbf{E}}(0)|^2 \rangle$. Starting from Eq. (6-29), we have

$$\langle |\hat{\mathbf{E}}(\tau)|^2 \, |(\hat{\mathbf{E}}(0)|^2 \rangle$$

$$= |\mathbf{A}|^4 \sum_{i,j} \sum_{l,m} \left\langle \exp[i\mathbf{k}_1 \cdot (\mathbf{r}_{i0} - \mathbf{r}_{j0} + \mathbf{r}_{l0} - \mathbf{r}_{m_0})] \right.$$

$$\times \exp\left\{ i\mathbf{k}_1 \cdot \int_0^\tau \mathbf{U}'[\mathbf{r}(t';\mathbf{r}_{i0}), t'] \right.$$

$$\left. \times \, dt' - i\mathbf{k}_1 \cdot \int_0^\tau \mathbf{U}'[\mathbf{r}(t';\mathbf{r}_{j0}), t']\, dt' \right\} \Big\rangle$$

$$= |\mathbf{A}|^4 N^2 + |\mathbf{A}|^4 N(N-1) \left\langle \exp\left\{ i\mathbf{k}_1 \cdot \int_0^\tau \mathbf{U}'[\mathbf{r}(t';\mathbf{r}_{i0}), t']\, dt' \right\} \right.$$

$$\times \exp\left\{ -i\mathbf{k}_1 \cdot \int_0^\tau \mathbf{U}'[\mathbf{r}(t';\mathbf{r}_{j0}), t']\, dt' \right\} \Big\rangle$$

$$+ |\mathbf{A}|^4 N(N-1) \left\langle \exp\left\{ i\mathbf{k}_1 \cdot \int_0^\tau \mathbf{U}'[\mathbf{r}(t';\mathbf{r}_{i0}), t']\, dt' \right\} \right.$$

$$\times \exp\left\{ -i\mathbf{k}_1 \cdot \int_0^\tau \mathbf{U}'[\mathbf{r}(t';\mathbf{r}_{j0}), t']\, dt' \right\}$$

$$\times \exp[2i\mathbf{k}_1 \cdot (\mathbf{r}_{i0} - \mathbf{r}_{j0})] \Big\rangle$$

$$+ |\mathbf{A}|^4 N^2(N-1) \left\langle \exp\left\{ i\mathbf{k}_1 \cdot \int_0^\tau \mathbf{U}'[\mathbf{r}(t';\mathbf{r}_{i0}), t']\, dt' \right\} \right.$$

$$\times \exp\left\{ -i\mathbf{k}_1 \cdot \int_0^\tau \mathbf{U}'[\mathbf{r}(t';\mathbf{r}_{j0}), t']\, dt' \right\} \exp[i\mathbf{k}_1 \cdot (\mathbf{r}_{i0} - \mathbf{r}_{j0})] \Big\rangle$$

$$(6\text{-}62)$$

where r_{i0} and r_{j0} are the initial positions of two general particles. The terms on the right-hand side of Eq. (6-62) correspond, respectively, to the cases (I) $i = j$, $m = l$; (II) $i = m \neq j = l$; (III) $i = l \neq j = m$; (IV) $i \neq j, m = l$, while the other terms vanish due to the presence of averages of oscillating factors of the type $\langle \exp(i\mathbf{k}_1 \cdot \mathbf{r}_{l0}) \rangle$, $\langle \exp(i\mathbf{k}_1 \cdot \mathbf{r}_{m0}) \rangle$.

Let us first take the average over the turbulent field. This requires evaluating the factor

$$\left\langle \exp\left\{ i\mathbf{k}_1 \cdot \int_0^\tau \mathbf{U}'[\mathbf{r}(t';\mathbf{r}_{i0}), t']\, dt' - i\mathbf{k}_1 \cdot \int_0^\tau \mathbf{U}'[\mathbf{r}(t';\mathbf{r}_{j0}), t']\, dt' \right\} \right\rangle_{\!\!M} \tag{6-63}$$

which can be done if one resorts to the joint Gaussian distribution hypothesis [see Eq. (6-55)] generalized to different fluid elements, that is, to different particles. It can easily be shown that this amounts to saying that the distribution of the displacements of the ith and jth scatterers is a joint Gaussian one, which implies

$$\left\langle \exp\left\{ i\mathbf{k}_1 \cdot \int_0^\tau \mathbf{U}'[\mathbf{r}(t';\mathbf{r}_{i0}), t']\, dt' - i\mathbf{k}_1 \cdot \int_0^\tau \mathbf{U}'[\mathbf{r}(t';\mathbf{r}_{j0}), t']\, dt' \right\} \right\rangle_{\!\!M}$$

$$= \exp\left\{ -k_1^2 \int_0^\tau \int_0^\tau \langle U_1[\mathbf{r}_i(t'), t']\, U_1[\mathbf{r}_i(t''), t''] \rangle_M\, dt'\, dt'' \right\}$$

$$\times \exp\left\{ k_1^2 \int_0^\tau \int_0^\tau \langle U_1[\mathbf{r}(t';\mathbf{r}_{i0}), t']\, U_1[\mathbf{r}(t'';\mathbf{r}_{j0}), t''] \rangle_M\, dt'\, dt'' \right\} \tag{6-64}$$

We could introduce at this point a Lagrangian correlation function $\tilde{R}_L$ of the velocities generalized to different fluid elements (Leslie, 1973), defined as

$$\langle U_1 [\mathbf{r}(t';\mathbf{r}_{i0}), t'] \, U_1 [\mathbf{r}(t'';\mathbf{r}_{j0}), t'']\rangle_M = \overline{U_1^2} \, \tilde{R}_L(t'' - t';\mathbf{r}_{i0} - \mathbf{r}_{j0})$$

(6-65)

which depends on $\mathbf{r}_{i0}$ and $\mathbf{r}_{j0}$ through the difference $\mathbf{r}_{i0} - \mathbf{r}_{j0}$ because of the homogeneity of the system.

It is now possible to show that the last two terms on the right-hand side of Eq. (6-62) are negligible with respect to the others. To this end, we rewrite Eq. (6-62), with the help of Eqs. (6-64) and (6-65), in the form

$$\langle |\hat{\mathbf{E}}(\tau)|^2 \, |\hat{\mathbf{E}}(0)|^2 \rangle$$

$$= |\mathbf{A}|^4 \, N^2 + |\mathbf{A}|^4 \, N(N-1)$$

$$\times \exp\left[-k_1^2 \, \overline{U_1^2} \int_0^\tau \int_0^\tau \tilde{R}_L(t'' - t') \, dt' \, dt'' \right]$$

$$\times \left\{ \left\langle \exp\left[k_1^2 \, \overline{U_1^2} \int_0^\tau \int_0^\tau \tilde{R}_L(t'' - t';\mathbf{r}_{i0} - \mathbf{r}_{j0}) \, dt' \, dt'' \right] \right\rangle_P \right.$$

$$+ N \left\langle \exp\left[k_1^2 \, \overline{U_1^2} \int_0^\tau \int_0^\tau \tilde{R}_L(t'' - t';\mathbf{r}_{i0} - \mathbf{r}_{j0}) \, dt' \, dt'' \right] \right.$$

$$\times \exp\left[i\mathbf{k}_i \cdot (\mathbf{r}_{i0} - \mathbf{r}_{j0}) \right] \Bigg\rangle_P$$

$$+ \left\langle \exp\left[k_1^2 \, \overline{U_1^2} \int_0^\tau \int_0^\tau \tilde{R}_L(t'' - t';\mathbf{r}_{i0} - \mathbf{r}_{j0}) \, dt' \, dt'' \right] \right.$$

$$\left. \times \exp\left[2i\mathbf{k}_1 \cdot (\mathbf{r}_{i0} - \mathbf{r}_{j0}) \right] \Bigg\rangle_P \right\}$$

(6-66)

In a significant experiment we could be limited to considering a time interval not larger than the inverse of the bandwidth $\Delta\omega$ of the scattered light, whose magnitude is approximately given by $k_1(\overline{U_1^2})^{1/2}$ [see Eq. (6-60)]. Since this time is, in practice, smaller than the correlation times for both R_L and $\tilde{R}_L$, we have

$$\langle |\hat{\mathbf{E}}(\tau)|^2 \, |\hat{\mathbf{E}}(0)|^2 \rangle$$

$$= |A|^4 \, N^2 + |A|^4 \, \frac{N(N-1)}{V_{sc}^2}$$

$$\times \left[\iint\limits_{V_{sc}} \exp\{-k_1^2 \, \overline{U_1^2} \, [1 - f(\mathbf{r}' - \mathbf{r}'')] \, \tau^2\} \, d\mathbf{r}' \, d\mathbf{r}'' \right.$$

$$+ \, N \iint\limits_{V_{sc}} \exp\{-k_1^2 \, \overline{U_1^2} \, [1 - f(\mathbf{r}' - \mathbf{r}'')] \, \tau^2\}$$

$$\times \exp[i\mathbf{k}_1 \cdot (\mathbf{r}' - \mathbf{r}'')] \, d\mathbf{r}' \, d\mathbf{r}''$$

$$+ \iint\limits_{V_{sc}} \exp\{-k_1^2 \, \overline{U_1^2} \, [1 - f(\mathbf{r}' - \mathbf{r}'')] \, \tau^2\}$$

$$\left. \times \exp[2i\mathbf{k}_1 \cdot (\mathbf{r}' - \mathbf{r}'')] \, d\mathbf{r}' \, d\mathbf{r}'' \right] \tag{6-67}$$

where $f(\mathbf{r}' - \mathbf{r}'') \equiv \tilde{R}_L(0; \mathbf{r}' - \mathbf{r}'')$ is a quantity whose magnitude ranges from unity to zero in a correlation length l_c. Accordingly, taking into account the relation $k_1^2 \, \overline{U_1^2} \, \tau^2 \lesssim 1$, the spatial variation of $\exp\{-k_1^2 \, \overline{U_1^2} \, [1 - f(\mathbf{r}' - \mathbf{r}'')] \, \tau^2\}$ takes place over a scale of the order $l_c \gg 1/k$, a fact that allows us to neglect the last two terms on the right-hand side of Eq. (6-67) due to the presence of the rapidly oscillating terms $\exp\{i\mathbf{k}_1 \cdot (\mathbf{r}' - \mathbf{r}'')\}$ and $\exp\{2i\mathbf{k}_1 \cdot (\mathbf{r}' - \mathbf{r}'')\}$.

VI.7 Non-Gaussian Nature of the Scattered Radiation

According to the results of the previous section, the equation the intensity correlation obeys is

$$
\langle |\hat{\mathbf{E}}(\tau)|^2 |\hat{\mathbf{E}}(0)|^2 \rangle
$$

$$
= |\mathbf{A}|^4 N^2 + |\mathbf{A}|^4 \frac{N(N-1)}{V_{sc}^2}
$$

$$
\times \iint_{V_{sc}} \exp\{-k_1^2 \overline{U_1^2}[1 - f(\mathbf{r}' - \mathbf{r}'')] \tau^2\} \, d\mathbf{r}' \, d\mathbf{r}'' \quad (6\text{-}68)
$$

which, in the limit $N \gg 1$, can be rewritten, according to Eq. (6-56), as (Di Porto *et al.*, 1969)

$$
\langle |\hat{\mathbf{E}}(\tau)|^2 |\hat{\mathbf{E}}(0)|^2 \rangle = \langle |\hat{\mathbf{E}}(0)|^2 \rangle^2 + |\langle \hat{\mathbf{E}}(\tau) \cdot \hat{\mathbf{E}}^*(0) \rangle|^2 \frac{1}{V_{sc}^2}
$$

$$
\times \iint_{V_{sc}} \exp[k_1^2 \overline{U_1^2} f(\mathbf{r}' - \mathbf{r}'') \tau^2] \, d\mathbf{r}' \, d\mathbf{r}'' \quad (6\text{-}69)
$$

The difference between Eq. (6-69) and the one expressing the usual Gaussian factorization lies in the presence of the factor

$$
F = \frac{1}{V_{sc}^2} \iint_{V_{sc}} \exp[k_1^2 \overline{U_1^2} f(\mathbf{r}' - \mathbf{r}'')] \, d\mathbf{r}' \, d\mathbf{r}'' \quad (6\text{-}70)
$$

In particular, in the limit $V_{sc} \gg l_c^3$, the exponential appearing in Eq. (6-70) can be set equal to unity except for a negligible region of the order l_c^3, so that $F \simeq 1$ and we recover the normal distribution formula as was expected on the basis of the central limit theorem. The influence of the factor F becomes increasingly relevant as V_{sc} approaches l_c^3. In the limit $V_{sc} \ll l_c^3$, $f(\mathbf{r}' - \mathbf{r}'')$ can be set equal to unity in Eq. (6-70) so that Eq. (6-69) yields, with the help of Eq. (6-56),

$$
\langle |\hat{\mathbf{E}}(\tau)|^2 |\hat{\mathbf{E}}(0)|^2 \rangle = 2 \langle |\hat{\mathbf{E}}(0)|^2 \rangle^2 \quad (6\text{-}71)
$$

Thus a measurement of the intensity correlation performed with different scattering volumes can give an estimate of the correlation length l_c, which is in turn connected with the dimension of the turbulent eddies. More explicitly, let us consider a first- and a second-order measurement, respectively, performed by means of the heterodyne and self-beating techniques, which give the optical spectrum $I(\omega)$ and the intensity fluctuation spectrum $P(\omega)$ [defined in Eq. (5-23)]. According to Eq. (6-60), $I(\omega)$ has a Gaussian form with bandwidth $\Delta\omega$ independent of the scattering volume, while this is not in general true for $P(\omega)$ as easily seen by inserting Eq. (6-69) into Eq. (5-23). (We do not consider for simplicity the δ-type contribution at frequency zero.) In the limit $V_{sc} \gg l_c^3$, however, one can apply Eq. (5-71), so that $P(\omega)$ has also a Gaussian shape with bandwidth $\sqrt{2}\,\Delta\omega$. As V_c is made to decrease, besides the departure of $P(\omega)$ from a Gaussian form (a circumstance not easy to verify experimentally), we observe that its bandwidth becomes smaller and tends to become zero in the limit $V_{sc} \ll l_c^3$.

This kind of measurement has been performed by Bourke *et al.* (1969, 1970) for studying turbulence associated with the flow of water in a pipe. They have actually been able to compare the bandwidth of $P(\omega)$, measured with the self-beating technique, with that of $I(\omega)$, measured with the heterodyne technique and independent of the scattering volume, thus obtaining an order of magnitude for the correlation length of the velocity field (see Figs. 6.11 and 6.12).

We wish now to show how, with a proper choice of scattering geometry, it is possible to obtain with this kind of measurement the detailed behavior of $f(\mathbf{r}' - \mathbf{r}'')$ (Bertolotti *et al.*, 1971).

This can be done, according to the geometry sketched in Fig. 6.13, by collecting on a single detector the radiation scattered at the same angle by two volumes V_1, V_2 much smaller than l_c^3 and separated by a distance ρ. Then the intensity fluctuation spectrum of the resulting field has a direct dependence on $f(\mathbf{\rho})$. This is readily seen by adapting Eq. (6-68) to the present

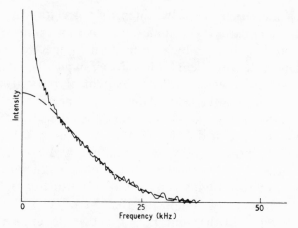

Fig. 6.11   *Self-beating spectrum of light scattered by particles suspended in a turbulent flow, in a situation in which $V_{sc} \gg l_c^3$. The profile is Gaussian and the bandwidth $\sqrt{2}$ times larger than the one pertaining to the curve of Fig. 6.10 (after Bourke et al., 1970).*

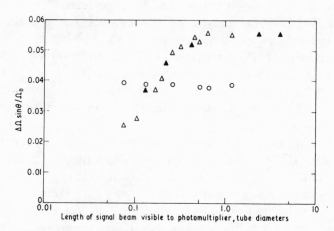

Fig. 6.12   *Linewidths of the power (Doppler) spectrum and of the intensity-fluctuation spectrum as a function of the ratio between the scattering volume and the correlation valume (after Bourke et al., 1970). Open circles correspond to the power spectrum, triangles to the intensity-fluctuation spectrum.*

situation thus obtaining

$$\langle |\hat{\mathbf{E}}(\tau)|^2 |\hat{\mathbf{E}}(0)|^2 \rangle = |\mathbf{A}|^4 N^2 + \frac{|\mathbf{A}|^4 N^2}{(V_1 + V_2)^2} [V_1^2 + V_2^2$$

$$+ 2V_1 V_2 \exp\{-k_1^2 \overline{U_1^2} [1 - f(\boldsymbol{\rho})] \tau^2\}] \quad (6\text{-}72)$$

In particular, by choosing $V_1 = V_2$ for simplicity, the intensity fluctuation spectrum $P(\omega)$ is, apart from the δ-function contributions at $\omega = 0$,

$$P(\omega) = \frac{|\mathbf{A}|^4 N^2}{4\pi} \left(\frac{c}{8\pi}\right)^2 \left[\frac{\pi}{k_1^2 \overline{U_1^2} [1 - f(\boldsymbol{\rho})]}\right]^{1/2}$$

$$\times \exp\left[\frac{\omega^2}{4k_1^2 \overline{U_1^2} [1 - f(\boldsymbol{\rho})]}\right] \quad (6\text{-}73)$$

so that $f(\boldsymbol{\rho})$ can be expressed in terms of its bandwidth $\Delta(\boldsymbol{\rho})$ by means of the formula

$$f(\boldsymbol{\rho}) = 1 - \frac{\Delta^2(\boldsymbol{\rho})}{\Delta^2(\infty)} \quad (6\text{-}74)$$

The behavior of $f(\boldsymbol{\rho})$ can then be obtained by measuring the bandwidth of $P(\omega)$ as a function of the distance ρ between the two scattering volumes.

We conclude these considerations on the non-Gaussian nature of the scattered field by noting that this is a direct consequence

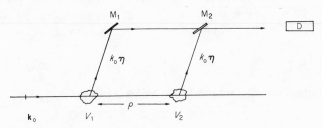

Fig. 6.13 *Schematic experimental arrangement for measuring the spatial behavior of the velocity correlation.*

of the correlation between two particles in the same vortex, so that the motions of the scatterers cannot be considered as practically independent whenever vortices of dimension comparable with the scattering volume exist.

In other words, turbulence furnishes a peculiar situation in which the correlation between the scatterers is not due to mutually exchanged forces but to a macroscopic stochastic motion independent of the presence of the scatterers themselves.

As for laminar regimes, light scattering by suspended indicator particles, already present or artificially added in a turbulent fluid, has become a standard technique for measuring the strength of turbulence [see, for example, Whitelaw, 1973].

Finally we wish to mention that the possibility of studying turbulence by observing the light scattered by an unseeded fluid, that is, by its dielectric constant fluctuations associated with hydrodynamical pressure fluctuations, has been investigated by Frisch (1967). This can be accomplished at least in principle by means of a first-order measurement, that is, by measuring the optical frequency spectrum at different angles, which turns out to be related to the velocity correlation function. While the ratio of the total cross section due to turbulence to that of usual thermal molecular scattering can be made larger than one for reasonable Reynolds numbers, in most practical situations this kind of experiment is difficult to realize due to the unavoidable presence of seeding particles. The method could be of practical usefulness when considering scattering by the upper atmosphere and troposphere as first proposed by Villars and Weisskopf (1954).

VI.8 Light Scattering from a Small Number of Particles

The situations considered in the previous sections are characterized by the presence of a large number N of particles in the scattering volume, so that the central limit theorem assures us of the Gaussian statistical behavior of the scattered radiation whenever the scatterers are independent. This in turn implies that an eventual departure from this kind of statistics is entirely due to the presence of a correlation in the motion of particles. In effect, the normal distribution hypothesis must be justified by giving an estimate of the non-Gaussian correction in terms of N. This has been explicitly considered (Chen and Tartaglia, 1972) for a system of a fixed number N of independent particles, and the correction turns out to be of the order $1/N$, in the sense that, for example, the moments of the scattered intensity obey the relation

$$\frac{\langle I^k \rangle}{\langle I \rangle^k} = k! + O(1/N), \qquad k = 1, 2, \ldots \qquad (6\text{-}75)$$

It is thus evident that whenever the scattering volume contains a small number of particles the corrections to the normal factorization property can become relevant. On the other hand, a small number of scatterers is often associated with a small scattering volume from which particles are free to escape, thus giving rise to fluctuations in their number. The theory of scattering by small fluctuating numbers of independent particles was first considered by Schaefer and Berne (1972). In order to describe this process, we consider a scattering volume V_{sc} contained in a much larger container of volume V filled with a fixed number of particles M and start from Eq. (6-29), suitably modified in the form

$$\hat{\mathbf{E}}(\mathbf{r}, t) = \mathbf{A} \sum_{i=1}^{M} b_i(t) \exp\{i[\mathbf{k}_1 \cdot \mathbf{r}_i(t) - \omega_0 t]\} \qquad (6\text{-}76)$$

where the factors $b_i(t)$, such that $b_i(t) = 1$ if the ith particle is in the scattering volume at time t and $b_i(t) = 0$ if it is not, take into account the possibility that the scatterers enter or leave V_{sc}. We can then write the first- and second-order correlation functions of the scattered field as

$$\langle \hat{\mathbf{E}}(t) \cdot \hat{\mathbf{E}}^*(0) \rangle = |\mathbf{A}|^2 \sum_{i=1}^{M} \sum_{j=1}^{M} \langle b_i(t) b_j(0)$$

$$\times \exp[i\mathbf{k}_1 \cdot \mathbf{r}_i(t) - i\mathbf{k}_1 \cdot \mathbf{r}_{j0}] \rangle e^{-i\omega t} \qquad (6\text{-}77)$$

$$\langle |\hat{\mathbf{E}}(t)|^2 |\hat{\mathbf{E}}(0)|^2 \rangle = |\mathbf{A}|^4 \sum_{i=1}^{M} \sum_{j=1}^{M} \sum_{m=1}^{M} \sum_{n=1}^{M} \langle b_i(t) b_j(t) b_m(0) b_n(0)$$

$$\times \exp[i\mathbf{k}_1 \cdot (\mathbf{r}_i(t) - \mathbf{r}_j(t) + \mathbf{r}_{m0} - \mathbf{r}_{n0})] \rangle \qquad (6\text{-}78)$$

If we consider particles undergoing a stochastic scattering and, as usual, scattering volumes satisfying the condition $V_{sc} \gg 1/k_1^3$, Eqs. (6-77) and (6-78) are superseded by

$$\langle \hat{\mathbf{E}}(t) \cdot \hat{\mathbf{E}}^*(0) \rangle = e^{-i\omega t} |\mathbf{A}|^2 \sum_{i=1}^{M} \sum_{j=1}^{M} \langle b_i(t) b_j(0) \rangle_S$$

$$\times \langle \exp\{i\mathbf{k}_1 \cdot [\mathbf{r}_i(t) - \mathbf{r}_{j0}]\} \rangle_F \qquad (6\text{-}79)$$

and

$$\langle |\hat{\mathbf{E}}(t)|^2 |\hat{\mathbf{E}}(0)|^2 \rangle = |\mathbf{A}|^4 \sum_{i=1}^{M} \sum_{j=1}^{M} \sum_{m=1}^{M} \sum_{n=1}^{M} \langle b_i(t) b_j(t) b_m(0) b_n(0) \rangle_S$$

$$\times \langle \exp\{i\mathbf{k}_1 \cdot [\mathbf{r}_i(t) - \mathbf{r}_j(t) + \mathbf{r}_{m0} - \mathbf{r}_{n0}]\} \rangle_F$$

$$(6\text{-}80)$$

where the factorization of the averages appearing in Eqs. (6-77) and (6-78) is a consequence of the large difference in the time scales of fluctuation of b_i and of $\exp(i\mathbf{k}_1 \cdot \mathbf{r}_i)$. As a matter of fact, the b_i's vary on a slow (S) scale of the order of the time spent by the particle in crossing the scattering volume, while the exponential factors oscillate on a fast (F) scale of the order of the time spent by the particle in moving a wavelength. Accordingly, the above factorization property pertains to a large class of scattering motions (turbulence, Brownian motion, and so on).

For statistically independent particles, Eqs. (6-79) and (6-80) immediately simplify to

$$\langle \hat{\mathbf{E}}(t) \cdot \hat{\mathbf{E}}^*(0) \rangle = e^{-i\omega_0 t} |\mathbf{A}|^2 \sum_{i=1}^{M} \langle b_i(t) b_i(0) \rangle_{\mathrm{S}}$$

$$\times \langle \exp\{ i\mathbf{k}_1 \cdot [\mathbf{r}_i(t) - \mathbf{r}_{i0}] \} \rangle_{\mathrm{F}} \qquad (6\text{-}81)$$

and

$$\langle |\hat{\mathbf{E}}(t)|^2 |\hat{\mathbf{E}}(0)|^2 \rangle = |\mathbf{A}|^4 \sum_{i,m=1}^{M} \langle b_i^2(t) b_m^2(0) \rangle_{\mathrm{S}} + |\mathbf{A}|^4 \sum_{i \neq m}^{M}$$

$$\times \langle b_i(t) b_i(0) \rangle_{\mathrm{S}} \langle b_m(t) b_m(0) \rangle_{\mathrm{S}}$$

$$\times \langle \exp\{ i\mathbf{k}_1 \cdot [\mathbf{r}_i(t) - \mathbf{r}_{i0}] \} \rangle_{\mathrm{F}}$$

$$\times \langle \exp\{ -i\mathbf{k}_1 \cdot [\mathbf{r}_m(t) - \mathbf{r}_{m0}] \} \rangle_{\mathrm{F}} \qquad (6\text{-}82)$$

We can now observe that the difference in the time scales allows for a good approximation setting $b_i(t) = b_i(0)$, $b_m(t) = b_m(0)$ in the right-hand side of Eq. (6-81) and in the second term in the right-hand side of Eq. (6-82), since they do not vary in the time interval necessary for the fast correlations to vanish. Furthermore, since $b_i^2(t) = b_i(t)$ and

$$\left\langle \sum_{i=1}^{M} b_i(t) \right\rangle_{\mathrm{S}} = M \langle b_i(t) \rangle_{\mathrm{S}} = \langle N(t) \rangle = M \frac{V_{\mathrm{sc}}}{V} \qquad (6\text{-}83)$$

where $N(t)$ is the number of particles present at time t inside V_{sc}, Eqs. (6-81) and (6-82) take the form

$$\langle \hat{\mathbf{E}}(t) \cdot \hat{\mathbf{E}}^*(0) \rangle = e^{-i\omega_0 t} |\mathbf{A}|^2 \langle N \rangle F(\mathbf{k}, t) \qquad (6\text{-}84)$$

and

$$\langle |\hat{\mathbf{E}}(t)|^2 |\hat{\mathbf{E}}(0)|^2 \rangle = |\mathbf{A}|^4 \langle \delta N(t) \delta N(0) \rangle_{\mathrm{S}}$$

$$+ |\mathbf{A}|^4 \langle N \rangle^2 \left[1 + \frac{M(M-1)}{M^2} |F(\mathbf{k}_1, t)|^2 \right] (6\text{-}85)$$

where $\delta N = N - \langle N \rangle$ and $F(\mathbf{k}_1, t) = \langle \exp\{ i\mathbf{k}_1 \cdot [\mathbf{r}_i(t) - \mathbf{r}_{i0}] \} \rangle_{\mathrm{F}}$.

In the limit $M \gg 1$, one recovers the usual Gaussian factorization formula but for the presence of the term $|A|^4 \langle \delta N(t) \, \delta N(0) \rangle_S$, that is,

$$\langle |\hat{\mathbf{E}}(t)|^2 \, |\hat{\mathbf{E}}(0)|^2 \rangle = \langle |\hat{\mathbf{E}}(t)|^2 \rangle^2 + |\langle \hat{\mathbf{E}}(t) \cdot \hat{\mathbf{E}}^*(0) \rangle|^2$$
$$+ \langle |\hat{\mathbf{E}}(t)|^2 \rangle \frac{\langle \delta N(t) \, \delta N(0) \rangle_S}{\langle N \rangle^2} \qquad (6\text{-}86)$$

The non-Gaussian contribution due to the fluctuation in the scatterer number (occupation number fluctuations) does not practically affect a heterodyne measurement due to its slow time dependence, while it does become evident by studying the long-time behavior of the intensity-correlation function. This in turn brings direct information on the probability after-effect factor $P_{ae}(t)$, defined as the probability that a particle inside V_{sc} at time zero will find itself outside it at time t (Chandrasekhar, 1943). In fact, we have immediately

$$\langle \delta N(t) \, \delta N(0) \rangle_S = \left\langle \sum_{i=1}^{M} \sum_{j=1}^{M} b_i(t) b_j(0) \right\rangle_S - \langle N \rangle^2$$

$$= \frac{M(M-1)}{M} \langle N \rangle^2 - \langle N \rangle^2 + M \langle b_i(t) b_i(0) \rangle_S$$

$$= \langle N \rangle [1 - P_{ae}(t)] - \frac{\langle N \rangle^2}{M} \qquad (6\text{-}87)$$

having taken into account that from the definition of $P_{ae}(t)$ it follows that

$$P_{ae}(t) = 1 - \langle b_i(t) b_i(0) \rangle \frac{V}{V_{sc}} \qquad (6\text{-}88)$$

Equation (6-87) shows that the non-Gaussian term is of the order $1/\langle N \rangle$ with respect to the Gaussian one, so that its significance increases for small scattering volumes and particle densities.

A general expression for the behavior of $P_{ae}(t)$ can be found, in the case of Brownian motion, in the literature (Chandrasekhar,

1943). It possesses a decay constant of the order of L^2/D, while
the decay time of the fast correlation is of the order of $1/Dk_1^2$
[see Eq. (6-50)], where L is the typical dimension of the scattering
volume.

An extension of Eq. (6-86) to include higher-order correlation
functions of the form $\langle|\hat{E}(t)|^{2k}|\hat{E}(0)|^{2k}\rangle$ has been considered by
Chen *et al.* (1973). A fairly complete treatment of the statistics
of the scattered intensity (considered at a given time) has been
given by Schaefer and Pusey (1973a, b), who show that the
intensity distribution $P_N(I)$ relative to light scattered by a fixed
number N of random scatterers obeys the relation

$$P_N(I) = \tfrac{1}{2}\int_0^\infty du\, uJ_0(u\sqrt{I})\{J_0(u)\}^N \qquad (6\text{-}89)$$

where J_0 is the Bessel function of order zero and the intensity
scattered by the single particle has been set equal to unity.
We must then average over the occupation number distribution.
In particular, in the thermodynamic limit in which $M \to \infty$ and
$V_{sc}/V \to 0$, this distribution turns out to be Poisson
(Chandrasekhar, 1943) so that

$$P_{\langle N\rangle}(I) = \sum_{N=0}^\infty \frac{e^{-\langle N\rangle}}{N!}\langle N\rangle^N P_N(I)$$

$$= \tfrac{1}{2}\int_0^\infty du\, uJ_0(u\sqrt{I})\exp\{\langle N\rangle[J_0(u)-1]\} \qquad (6\text{-}90)$$

where $\langle N\rangle$ denotes the (finite) average number of scatterers.
It can be shown that Eqs. (6-89) and (6-90) both tend toward
the exponential-intensity distribution relative to a Gaussian field
(see Mandel and Wolf, 1965), respectively, in the limit $N \to \infty$
and $\langle N\rangle \to \infty$, as a priori expected from the central limit
theorem.

Experiments concerning the possibility of observing the
presence and the time behavior of the non-Gaussian contribution
have recently been performed by Schaefer and Pusey (1972) and

Koppel and Schaefer (1973). These kinds of measurements can be exploited for biological investigations concerning the dynamics of motile microorganisms (Schaefer, 1973).

VI.9 Further Examples of Non-Gaussian Scattered Light: Scattering by a Rough Surface and a Random-Phase Screen

Recent applications concerning the deviation of the statistics of scattered light from those of Gaussian light have been considered in the cases of an electromagnetic field scattered by a rough surface and, more generally, diffracted by a random-phase screen (Jakeman and Pusey, 1973; Jakeman, 1974).

In the first situation, we are dealing with a rough reflecting surface composed of many identical facets of characteristic linear dimension ξ much larger than the wavelength λ of the incident beam. The orientations of the facets are mutually independent random functions of time, this being the source of the randomness of the back-scattered field observed in the far-field zone by means of a point detector. The assumption $\xi/\lambda \gg 1$ ensures that the light incident on each facet is reflected in a direction determined by the laws of geometrical optics, with an angular uncertainty associated with the first diffraction lobe corresponding to a small solid angle $\delta\Omega \simeq (\lambda/\xi)^2$. The reflecting surface is supposed to lie on the average in a fixed plane, and the mean square root of its random inclination over this plane is assumed to be much larger than λ/ξ, which corresponds to the relation $(\overline{Z^2})^{1/2} \gg \lambda$ (see Fig. 6.14). This last hypothesis implies that the zone scanned by the reflected rays greatly exceeds their angular uncertainty $\delta\Omega$, so that each facet can be considered as an *effective* scatterer in a given direction for only a small fraction of time.

In the situation described, we can write the expression of the field reflected in a given direction η in a form analogous to that of

Eq. (6-76), that is,

$$\hat{E}(\eta, t) = A \sum_{i=1}^{M} b_i(\eta, t) \exp i[\varphi_i(t) - \omega_0 t] \qquad (6\text{-}91)$$

where M is the total number of facets illuminated by the incident beam, and the φ_i's are mutually independent random real quantities, as are the b_i's, which obey the relation $b_i = 1$ or $b_i = 0$, according to whether or not the light reflected by the ith facet impinges on the detector. In practice, it is reasonable to assume that, in the characteristic time τ_c of the angular fluctuations necessary for a facet to reflect twice the light in the same direction, the average translation length undergone by the facets is much larger than λ, which in particular is sufficient to assure the absence of correlations between $b_i(t)$ and $\varphi_i(t)$. We can then start from Eqs. (6-81) and (6-82) with the substitution $\mathbf{k}_1 \cdot \mathbf{r}_i(t) \rightarrow \varphi_i(t)$, and obtain immediately

$$\langle |\hat{E}(\eta, t)|^2 \rangle = |A|^2 \langle N \rangle \qquad (6\text{-}92)$$

and

$$\langle |\hat{E}(\eta, t)|^4 \rangle = |A|^4 \langle N \rangle + 2|A|^4 [M(M-1)/M^2] \langle N \rangle^2$$

$$(6\text{-}93)$$

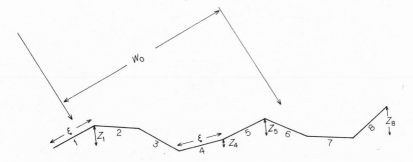

Fig. 6.14 *Schematic of a rough surface: the facets $1, 2, \ldots$ have a common linear dimension ξ and different inclination angles α_i, with respect to the average position of the surface, determined by the relation $\sin \alpha_i = Z_i/\xi$.*

where $\langle N \rangle$ is the average number of effective scattering facets and the relation $\langle b^2 \rangle = \langle b \rangle = \langle N \rangle / M$ has been used.

From Eqs. (6-92) and (6-93), we have

$$\frac{\langle |\hat{\mathbf{E}}(\boldsymbol{\eta}, t)|^4 \rangle - \langle |\hat{\mathbf{E}}(\boldsymbol{\eta}, t)|^2 \rangle^2}{\langle |\hat{\mathbf{E}}(\boldsymbol{\eta}, t)|^2 \rangle^2} = 1 - \frac{2}{M} + \frac{1}{\langle N \rangle} \qquad (6\text{-}94)$$

which shows a departure from Gaussian behavior depending on M and $\langle N \rangle$. Since $\langle N \rangle \simeq M(\lambda/\xi)^2 \ll M$, the non-Gaussian nature of the scattered radiation is directly connected to the number of facets possessing the right orientation and not to the total number of independent illuminated scatterers. More precisely, since

$$\langle N \rangle = MP(\Omega)\,\delta\Omega \qquad (6\text{-}95)$$

where $P(\Omega)$ represents the probability distribution of the facet orientation, by measuring the angular dependence of the non-Gaussian contribution in Eq. (6-94), we can in principle determine both $P(\Omega)$ and the characteristic facet size ξ.

By starting from Eqs. (6-81) and (6-82) with the substitution $\mathbf{k}_1 \cdot \mathbf{r}_i(t) \to \varphi_i(t)$, it is also possible to derive, after some straightforward algebra, the different time-normalized intensity correlation, which reads

$$\frac{\langle |\hat{\mathbf{E}}(\boldsymbol{\eta}, \tau)|^2 |\hat{\mathbf{E}}(\boldsymbol{\eta}, 0)|^2 \rangle}{\langle |\hat{\mathbf{E}}(\boldsymbol{\eta}, \tau)|^2 \rangle^2}$$

$$= \left(1 - \frac{1}{M}\right)\left(1 + \frac{|\langle \hat{\mathbf{E}}(\boldsymbol{\eta}, \tau) \cdot \hat{\mathbf{E}}^*(\boldsymbol{\eta}, 0) \rangle|^2}{\langle |\hat{\mathbf{E}}(\boldsymbol{\eta}, \tau)|^2 \rangle^2}\right) + M \frac{\langle b(\tau) b(0) \rangle}{\langle N \rangle^2}$$

$$(6\text{-}96)$$

where b indicates a generic b_i. The temporal behavior of Eq. (6-96) is characterized by the time τ_c associated with the orientational fluctuation of the facets and the time $\tau_c' = \tau_c / [k(\overline{Z^2})^{1/2}]$, which represents the inverse of the typical Doppler frequency shift suffered by the reflected electric field $[\tau_c' \simeq (kv)^{-1}$, where v is an average velocity of the scatterers given by $v = (\overline{Z^2})^{1/2}/\tau_c$]. More precisely, while the Gaussian term in the right-hand side of Eq. (6-96) decays to a constant value in the fast time τ'_c, since after this time the cross correlation of the scattered field vanishes, the behavior of the non-Gaussian term will in general exhibit both characteristic times τ_c and τ'_c.

The fluctuating rough surface can be considered as a particular case of the random-phase screen. This is a system that induces phase variations in a propagating electromagnetic wave, so that there is a surface over which the emerging field presents phase fluctuations in space and time. The light diffracted from this field distribution turns out to possess non-Gaussian statistical properties.

The electromagnetic field on the screen can be written as

$$\hat{\mathbf{E}}(\mathbf{r}, t) = \mathbf{E}_0 \exp\{i[\Phi(\mathbf{r}, t) - \omega_0 t] - r^2/W_0^2\} \qquad (6\text{-}97)$$

where $\mathbf{r}$ represents the position on the plane of the screen and an incident Gaussian intensity profile of the type

$$I = I_0 \exp(-2r^2/W_0^2)$$

has been assumed. Besides, we assume that W_0 is large enough and the phase fluctuations small enough so that Eq. (6-97) is consistent, in first approximation, with a perturbation propagating in the z-direction orthogonal to $\mathbf{r}$. Under these conditions, the Kirchhoff–Huygens diffraction integral can be applied with good approximation in the form valid for a plane wave of wave number $k_0 = \omega_0/c$, thus obtaining the diffracted field at position $\mathbf{R}$ as (see, for example, Marcuse, 1972)

$$\hat{\mathbf{E}}(\mathbf{R}, t) = \mathbf{E}_0 \frac{k_0 \exp(-i\omega_0 t)}{4\pi i} \int d\mathbf{r} \, (1 + \cos \vartheta)$$

$$\times \exp[i\Phi(\mathbf{r}, t) - r^2/W_0^2] \frac{\exp(ik_0|\mathbf{R} - \mathbf{r}|)}{|\mathbf{R} - \mathbf{r}|}$$

$$(6\text{-}98)$$

where ϑ is the angle between $\mathbf{R} - \mathbf{r}$ and the z-direction. At observation distances much larger than W_0, we have approximately

$$\hat{\mathbf{E}}(\mathbf{R}, t) = \frac{k_0 \mathbf{E}_0}{4\pi i R} \exp(-i\omega_0 t)(1 + \cos \vartheta) \int d\mathbf{r} \exp[i\Phi(\mathbf{r}, t) - r^2/W_0^2]$$

$$\times \exp[ik_0 R(1 - r\sin\vartheta\cos\psi/R)] \qquad (6\text{-}99)$$

where ϑ is now a well-defined scattering angle and ψ the polar angle of integration corresponding to $\mathbf{r}$, having chosen as polar axis the projection of $\mathbf{R}$ on the screen.

We can now evaluate the first- and second-order correlation functions of the diffracted field. From Eq. (6-99),

$$\langle \hat{\mathbf{E}}(\mathbf{R}, t+\tau) \cdot \hat{\mathbf{E}}^*(\mathbf{R}, t) \rangle = [k_0^2 |\mathbf{E}_0|^2/(4\pi R)^2](1 + \cos\vartheta)^2$$

$$\times \exp(-i\omega_0 \tau) \int d\mathbf{r}'\, d\mathbf{r}''$$

$$\times \langle \exp\{i[\Phi(\mathbf{r}', t+\tau) - \Phi(\mathbf{r}'', t)]\} \rangle$$

$$\times \exp[ik_0 \sin\vartheta(r''\cos\psi'' - r'\cos\psi')]$$

$$\times \exp[-(r'^2 + r''^2)/W_0^2] \qquad (6\text{-}100)$$

A joint Gaussian distribution hypothesis for the $\Phi(\mathbf{r}_i, t_i)$'s implies, as an easy generalization of Eq. (6-48),

$$\langle \exp[i\Phi(\mathbf{r}', t+\tau) - i\Phi(\mathbf{r}'', t)] \rangle = \exp\{-\overline{\Phi^2}[1 - \rho(|\mathbf{r}' - \mathbf{r}''|)\sigma(\tau)]\} \qquad (6\text{-}101)$$

where a phase correlation function has been assumed, of the type

$$\langle \Phi(\mathbf{r}', t+\tau)\Phi(\mathbf{r}'', t) \rangle = \overline{\Phi^2}\,\rho(|\mathbf{r}' - \mathbf{r}''|)\sigma(\tau) \qquad (6\text{-}102)$$

with $\rho(0) = \sigma(0) = 1$. After straightforward algebra, Eq. (6-100) can be rewritten, with the help of Eqs. (6-101) and (6-102), as

$$\langle \hat{\mathbf{E}}(\mathbf{R}, t+\tau) \cdot \hat{\mathbf{E}}^*(\mathbf{R}, t) \rangle$$

$$= \frac{k_0^2 |\mathbf{E}_0|^2 W_0^2}{16R^2} (1 + \cos \vartheta)^2 \exp(-i\omega_0 \tau)$$

$$\times \int_0^\infty J_0(k_0 r \sin \vartheta) \exp\{-\overline{\Phi^2}[1 - \rho(r)\sigma(\tau)]\}$$

$$\times \exp(-r^2/2W_0^2) r \, dr \tag{6-103}$$

where J_0 is the zeroth order Bessel's function.

Equation (6-103) can be cast in a simpler form if we observe that, in many situations, $\overline{\Phi^2} \gg 1$, so that only small values of r are important in the exponential appearing in the right-hand side of Eq. (6-103). By writing approximately

$$\rho(r) = 1 - r^2/\xi^2, \quad \text{for} \quad r/\xi \ll 1 \tag{6-104}$$

it is possible to perform the integral appearing in Eq. (6-103), obtaining for all significant times

$$\langle \hat{\mathbf{E}}(\mathbf{R}, t+\tau) \cdot \hat{\mathbf{E}}^*(\mathbf{R}, t) \rangle = \frac{k_0^2 |\mathbf{E}_0|^2 W_0^2}{16R^2} (1 + \cos \vartheta)^2$$

$$\times \exp(-i\omega_0 \tau) \exp\{\overline{\Phi^2}[\sigma(\tau) - 1]\}$$

$$\times \frac{\xi^2}{2\overline{\Phi^2}\sigma(\tau)}$$

$$\times \exp\{-k^2 \xi^2 \sin^2 \vartheta / [4\overline{\Phi^2} \sigma(\tau)]\} \tag{6-105}$$

where the reasonable hypothesis

$$\overline{\Phi^2} \sigma(\tau)/\xi^2 \gg 1/W_0^2 \tag{6-106}$$

has been used. By means of an analogous procedure, we can evaluate the second-order correlation function of the diffracted field, thus obtaining (Jakeman, 1974)

$$\frac{\langle |\hat{\mathbf{E}}(\mathbf{R}, t+\tau)|^2 |\hat{\mathbf{E}}(\mathbf{R}, t)|^2 \rangle}{\langle |\hat{\mathbf{E}}(\mathbf{R}, t)|^2 \rangle^2} = \left(1 - \frac{\xi^2}{W_0^2}\right)$$

$$\times \left[1 + \frac{|\langle \hat{\mathbf{E}}(\mathbf{R}, t+\tau) \cdot \hat{\mathbf{E}}^*(\mathbf{R}, t)\rangle|^2}{\langle |\hat{\mathbf{E}}(\mathbf{R}, t)|^2 \rangle^2}\right]$$

$$+ \frac{\xi^2 \overline{\Phi^2}}{W_0^2 \{4 + \overline{\Phi^2}[1 - \sigma(\tau)]\}}$$

$$\times \exp\left\{\frac{k^2 \xi^2 \sigma(\tau) \sin^2 \vartheta}{2\overline{\Phi^2}[1 + \sigma(\tau)]}\right\} \qquad (6\text{-}107)$$

Equations (6-105) and (6-107) allow us to relate the power and intensity fluctuation spectra of the diffracted light to the behavior of the phase correlation function. In particular, the decay time of the first-order correlation function can be shown to be significantly reduced with respect to the phase fluctuation time by the presence of a large factor related to $\overline{\Phi^2}$. Conversely, the non-Gaussian term of Eq. (6-107) exhibits, beyond this fast decay, a larger characteristic time of the order of the phase fluctuation time. At time $\tau = 0$, Eqs. (6-105) and (6-107) yield

$$\langle |\hat{\mathbf{E}}(\mathbf{R}, t)|^2 \rangle = \frac{k_0^2 |\mathbf{E}_0|^2 W_0^2 \xi^2}{32 R^2 \overline{\Phi^2}} (1 + \cos \vartheta)^2 \exp\left\{\frac{-k^2 \xi^2 \sin^2 \vartheta}{4\overline{\Phi^2}}\right\}$$

$$(6\text{-}108)$$

$$\frac{\langle |\hat{\mathbf{E}}(\mathbf{R}, t)|^4 \rangle}{\langle |\hat{\mathbf{E}}(\mathbf{R}, t)|^2 \rangle^2} = 2(1 - \xi^2/W_0^2) + \frac{\xi^2 \overline{\Phi^2}}{4 W_0^2} \exp\left\{\frac{k^2 \xi^2 \sin^2 \vartheta}{4\overline{\Phi^2}}\right\}$$

$$(6\text{-}109)$$

which show how an investigation of the diffracted intensity as a function of ϑ, and of the normalized mean square intensity as a function of W_0, allows us to determine both significant parameters $\overline{\Phi^2}$ and ξ. In particular, we note that the non-Gaussian regime is in practice observable by focalizing the beam to a dimension $W_0 \simeq \xi(\overline{\Phi^2})^{1/2}$, much larger than ξ.

The theoretical behavior predicted by Eqs. (6-108) and (6-109) has been experimentally verified (Jakeman and Pusey, 1973). The random-phase screen is furnished in this case by a thin layer of a nematic liquid crystal undergoing electrohydrodynamic turbulence, a situation in which a light beam may travel suffering mainly phase fluctuations (Deutsch and Keating, 1969). The experimental results are in good agreement with theory (see Figs. 6.15 and 6.16), which is also consistent with the data obtained in the same situation for the bandwidth of the spectrum of diffracted light (Scudieri *et al.*, 1974). Another relevant example of the random-phase screen is furnished by rotating ground

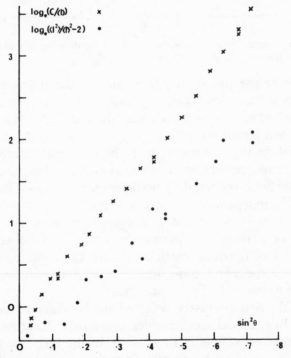

Fig. 6.15 *Angular dependence of* $\langle I \rangle \propto \langle |\hat{\mathbf{E}}(\mathbf{R}, t)|^2 \rangle$ *and* $\langle I^2 \rangle / \langle I \rangle^2 = \langle |\hat{\mathbf{E}}(\mathbf{R}, t)|^4 \rangle / \langle |\hat{\mathbf{E}}(\mathbf{R}, t)|^2 \rangle^2$, *where C is a convenient normalization constant (after Jakeman and Pusey, 1973b).*

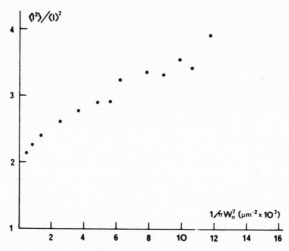

Fig. 6.16 *Dependence of* $\langle I^2 \rangle$ *on the size* W_0 *of the illuminated area at a fixed angle (after Jakeman and Pusey, 1973b).*

glass, where the non-Gaussian nature of the diffracted light has already been experimentally tested and interpreted in the framework of scattering by few scatterers (Bluemel *et al.*, 1972; Scudieri and Bertolotti, 1974).

We wish finally to emphasize the fundamental difference between the two types of non-Gaussian fields treated in the chapter, that is, the fields scattered by particles embedded in a turbulent fluid and by independent scatterers, respectively. In the first case, the central limit theorem does not apply, in spite of the large number of particles, whenever the correlation between the trajectories of different particles cannot be neglected. In the second case, the central limit theorem does not apply because of the small number of effective scatterers.

This fact is particularly reflected in the different behavior that the normalized second-order correlation function of the scattered fields exhibits at time $\tau = 0$. In the first case it takes on the value two [see Eq. (6-69)], since the central limit theorem can be applied at equal times, while in the second case this is no longer true [see Eqs. (6-86), (6-96), and (6-109)].

References

Abbiss, J. B., Chubb, T., and Pike, E. R. (1974). *Opt. and Laser Technol.* **6** (6).

Agarwal, G. S. (1970). *Phys. Rev.* **A2**, 2083.

Arecchi, F. T. (1965). *Phys. Rev. Lett.* **15**, 912.

Arecchi, F. T., Giglio, M., and Tartari, U. (1967). *Phys. Rev.* **163**, 186.

Becker, R. (1933). "Theorie der Elektrizität," Vol. II, 6th ed. Teubner, Leipzig.

Bedeaux, D., and Mazur, P. (1973). *Physica* **67**, 23.

Benedek, G. B. (1969). *In* "Polarisation, Matière et Rayonnement." Presse Univ. de France, Paris.

Beran, M. J., De Velis, J., and Parrent, G. B. (1967). *Phys. Rev.* **154**, 1224.

Bertolotti, M., Crosignani, B., Di Porto, P., and Sette, D. (1967). *Phys. Rev.* **157**, 146.

Bertolotti, M., Crosignani, B., Di Porto, P., and Sette, D. (1969). *J. Phys. A (Gen. Phys.)* **2**, 473.

Bertolotti, M., Crosignani, B., and Di Porto, P. (1970). *J. Phys. A. (Gen. Phys.)* **3**, L37.

Bertolotti, M., Crosignani, B., Daino, B., and Di Porto, P. (1971). *J. Phys. A (Gen. Phys.)* **4**, L47.

Bluemel, V., Narducci, L. M., and Tuft, R. A. (1972). *J. Opt. Soc. Am.* **62**, 1309.

Bonifacio, R., and Preparata, G. (1970). *Phys. Rev.* **A2**, 336.

Born, M., and Wolf, E. (1970). "Principles of Optics." Pergamon, Oxford.

Bourke, P. J., *et al.* (1969). *Phys. Lett.* **28A**, 692.

Bourke, P. J., *et al.* (1970). *J. Phys. A* (*Gen. Phys.*) **3**, 216.

Brillouin, L. (1922). *Ann. Phys.* (*Paris*) **17**, 88.

Bucaro, J. A., and Litovitz, T. A. (1971). *J. Chem. Phys.* **54**, 3846.

Bullough, R. K. (1967). *Proc. Interdisciplinary Conf. Electromagn. Scattering, 2nd, 1965* (R. L. Rowell and R. S. Stein, eds.). Gordon and Breach, New York.

Bullough, R. K. (1968). *J. Phys. A* (*Gen. Phys.*) **1**, 409.

Bullough, R. K., and Hynne, F. (1968). *Chem. Phys. Lett.* **2**, 307.

Bullough, R. K., Obada, A.-S. F., Thompson, B. V., and Hynne, F. (1968). *Chem. Phys. Lett.* **5**, 293.

Cantrell, C. D. (1968). Ph.D. Thesis, Princeton Univ., Princeton, New Jersey (unpublished).

Chandrasekhar, S. (1943). *Rev. Mod. Phys.* **15**, 1.

Chandrasekhar, S. (1951). *Proc. Roy. Soc.* **A207**, 301.

Chen, S. H., and Tartaglia, P. (1972). *Opt. Commun.* **6**, 119.

Chen, S. H., Tartaglia, P., and Pusey, P. N. (1973). *J. Phys. A* (*Gen. Phys.*) **6**, 490.

Chow, T. S. (1973). *Phys. Fluids*, **16**, 31.

Choy, T. R., and Mayer, J. E. (1967). *J. Chem. Phys.* **46**, 110.

Chu, B. (1970). *Ann. Rev. Phys. Chem.* **21**, 145.

Chu, B. (1974). "Laser Light Scattering." Academic Press, New York.

Clark, N. L., Lunacek, J. H., and Benedek, G. B. (1970). *Amer. J. Phys.* **38**, 575.

Cohen, C., Sutherland, J. W. H., and Deutch, J. M. (1971). *Phys. Chem. Liquids* **2**, 213.

Crosignani, B., and Di Porto, P. (1967). *Phys. Lett.* **24A**, 69.

Crosignani, B., Di Porto, P., and Engelmann, F. (1968a). *Z. Naturforsch.* **23a**, 743.

Crosignani, B., Di Porto, P., and Engelmann, F. (1968b). *Z. Naturforsch.* **23a**, 968.

Cummins, H. Z. (1974). *In* "Photon Counting and Light Beating Spectroscopy" (H. Z. Cummins and E. R. Pike, eds.) Plenum Press, New York.

Cummins, H. Z., and Pike, E. R. (eds.) (1974). "Photon Correlation and Light Beating Spectroscopy" (*Proc. NATO Advan. Study Inst., Capri, Italy, 1973*). Plenum Press, New York.

Cummins, H. Z., and Swinney, H. L. (1970). *In* "Progress in Optics" (E. Wolf, ed.), Vol. VIII. North-Holland Publ., Amsterdam.

Cummins, H. Z., Knable, N., and Yeh, Y. (1964). *Phys. Rev. Lett.* **4**, 176.

Cummins, H. Z., Carlson, F. D., Herbert, T. J., and Woods, G. (1969). *Biophys. J.* **9**, 518.

Darwin, C. G. (1924). *Trans. Cambridge Phys. Soc.* **23**, 137.

De Groot, S. R., and Mazur, P. (1962). "Non-Equilibrium Thermodynamics." North-Holland Publ., Amsterdam.

Deutsch, C., and Keating, P. N. (1969). *J. Appl. Phys.* **40**, 4049.

Di Porto, P., Crosignani, B., and Bertolotti, M. (1969). *J. Appl. Phys.* **40**, 5083.

Dubin, S. B., Lunacek, J. H., and Benedek, G. B. (1967). *Proc. Nat. Acad. Sci. U.S.* **57**, 1164.

Durst, F., Melling, A., and Whitelaw, J. H. (1972). *J. Fluid. Mech.* **56**, 143.

Einstein, A. (1910). *Ann. Phys.* **33**, 1275.

Eyring, H., Walter, J., and Kimball, G. E. (1944). "Quantum Chemistry." Wiley, New York.

Fabelinskii, I. L. (1968). "Molecular Scattering of Light." Plenum Press, New York.

Fetter, A. L. (1965). *Phys. Rev.* **139**, A1616.

Fidone, I., Lafleur, S., and Lafleur, C. (1963). Rep. EUR-CEA-FC No. 204, Fontenay-aux-Roses, Haute-de-Seine, France.

Fisher, I. Z. (1964). "Statistical Theory of Liquids." Chicago Univ. Press, Chicago, Illinois.

Fixman, M. (1955). *J. Chem. Phys.* **23**, 2074.

Fixman, M. (1960). *J. Chem. Phys.* **33**, 1357.

Fleury, P. A., and Boon, J. P. (1973). *In* "Advances in Chemical Physics" (I. Prigogine and S. A. Rice, eds.), Vol. 24. Wiley, New York.

Foch, J. (1968). *Phys. Fluids* **11**, 2336.

Ford, N. C., Jr., and Benedek, G. B. (1965). *Phys. Rev. Lett.* **15**, 649.

Foreman, J. W., George, E. W., and Lewis, R. D. (1965). *Appl. Phys. Lett.* **7**, 77.

Foreman, J. W., George, E. W., Jetton, J. L., Lewis, R. D., Thorton, J. R., and Watson, H. J. (1966). *IEEE J. Quantum Electron.* **QE-2**, 260.

Forrester, A. T., Gudmunsen, R. A., and Johnson, P. O. (1955). *Phys. Rev.* **99**, 1691.

Freed, C., and Haus, H. A. (1966). *In* "Physics of Quantum Electronics" (P. L. Kelley, B. Lax, and P. E. Tannenwald, eds.). McGraw-Hill, New York.

Frisch, H. L. (1967). *Phys. Rev. Lett.* **19**, 1278.

Fujima, S. (1969). *J. Phys. Soc. Japan* **27**, 1370.

Gelbart, W. M. (1972). *J. Chem. Phys.* **57**, 699.

Gelbart, W. M. (1974). *In* "Advances in Chemical Physics" (I. Prigogine and S. A. Rice, eds.), Vol. 26. Wiley, New York.

Gershon, N. D., and Oppenheim, I. (1973). *Physica* **64**, 247.

Giglio, M., and Benedek, G. B. (1969). *Phys. Rev. Lett.* **23**, 1145.

Glauber, R. J. (1952). *Phys. Rev.* **87**, 189 (abstract).

Glauber, R. J. (1954). *Phys. Rev.* **94**, 751 (abstract).

Glauber, R. J. (1962). *In* "Lectures in Theoretical Physics" (W. E. Brittin, B. W. Downs, and J. Downs, eds.), Vol. IV. Wiley (Interscience), New York.

Glauber, R. J. (1963a). *Phys. Rev.* **130**, 2529.

Glauber, R. J. (1963b). *Phys. Rev.* **131**, 2766.

Glauber, R. J. (1965). *In* "Quantum Optics and Electronics" (C. De Witt, A. Blandin, and C. Cohen-Tannoudji, eds.). Gordon and Breach, New York.

Glauber, R. J. (1966). *In* "Physics of Quantum Electronics" (P. L. Kelley, B. Lax, and P. E. Tannenwald, eds.). McGraw-Hill, New York.

Glauber, R. J. (1967). *In Symp. Mod. Opt. 1967.* Polytechnic Press, New York.

Glauber, R. J. (1969). *In Int. School Phys. Enrico Fermi, Varenna, 1967.* Academic Press, New York.

Glauber, R. J. (1970). *In* "Quantum Optics" (S. M. Kay and A. Maitland, eds., Scottish Univ. Summer School of Physics). Academic Press, New York.

Goldberger, M. L., Lewis, H. W., and Watson, K. M. (1963). *Phys. Rev.* **132**, 2764.

Goldstein, R. J., and Hagen, W. F. (1967). *Phys. Fluids* **10**, 1349.

Grad, H. (1958). *In* "Handbuch der Physik," Vol. XII. Springer-Verlag, Berlin.

Greytak, T. J., and Benedek, G. B. (1966). *Phys. Rev. Lett.* **17**, 179.

Gross, E. (1930). *Nature (London)* **126**, 201.

Haken, H. (1970). *In* "Handbuch der Physik," Vol. XXV/2c. Springer-Verlag, Berlin.

Hanbury Brown, R., and Twiss, P. Q. (1956). *Nature (London)* **177**, 27; **178**, 1046.

Hanbury Brown, R., and Twiss, P. Q. (1957). *Proc. Roy. Soc. A* **242**, 300.

Hellwarth, B. W. (1970). *J. Chem. Phys.* **52**, 2128.

Hinze, O. (1959). "Turbulence." McGraw-Hill, New York.

Hoek, H. (1941). *Physica* **8**, 209.

Hunt, F. V. (1957). *In* "American Institue of Physics Handbook" (D. E. Gray, ed.). McGraw-Hill, New York.

Jakeman, E. (1974). *In* "Photon Counting and Light Beating Spectroscopy" (H. Z. Cummins and E. R. Pike, eds.). Plenum Press, New York.

Jakeman, E., and Pusey, P. N. (1973a). *J. Phys. A (Math. Nucl. Gen.)* **6**, L88.

Jakeman, E., and Pusey, P. N. (1973b). *Phys. Lett.* **44A**, 456.

Jakeman, E., Oliver, C. J., and Pike, E. R. (1968). *J. Phys. A (Proc. Phys. Soc.)* **1**, 406.

Johnson, F. A., McLean, T. P., and Pike, E. R. (1966). *In* "Physics of Quantum Electronics" (P. L. Kelley, B. Lax, and P. E. Tannenwald, eds.). McGraw-Hill, New York.

Kadanoff, L. P., and Swift, J. (1968). *Phys. Rev.* **166**, 89.

Kawasaki, K. (1971). *In* "Critical Phenomena" (M. S. Green, ed.). Academic Press, New York.

Kelley, P. L., and Kleiner, W. H. (1964). *Phys. Rev.* **136**, A316.

Kerker, M. (1969). "The Scattering of Light." Academic Press, New York.

Klauder, J. R., and Sudarshan, E. C. G. (1968). "Fundamentals of Quantum Optics." Benjamin, New York.

Klimontovich, Yu. L. (1967). "The Statistical Theory of Non-Equilibrium Processes in Plasma." Pergamon, Oxford.

Komarov, L. I., and Fisher, I. Z. (1962). *Zh. Esperim. Teor. Fiz.* **43**, 1927 [*English Transl.: Sov. Phys.-JETP* **16**, 1358 (1963)].

Koppel, D. E., and Schaefer, D. W. (1973). *Appl. Phys. Lett.* **22**, 36.

Korenman, V. (1967). *Phys. Rev.* **154**, 1233.

Korenman, V. (1970). *Phys. Rev.* **A2**, 449.

Lallemand, P. (1974). *In* "Photon Counting and Light Beating Spectroscopy" (H. Z. Cummins and E. R. Pike, eds.). Plenum Press, New York.

Landau, L. D., and Lifshitz, E. M. (1958). "Statistical Physics." Pergamon, Oxford.

Landau, L. D., and Lifshitz, E. M. (1959). "Fluid Mechanics." Pergamon, Oxford.

Landau, L. D., and Lifshitz, E. M. (1960). "Electrodynamics of Continuous Media." Pergamon, Oxford.

Landau, L. D., and Placzek, G. (1934). *Phys. Z. Sovietun.* **5**, 172.

Lastovka, J. B., and Benedek, G. B. (1966). *Phys. Rev. Lett.* **17**, 1039.

Lebowitz, J. L., and Percus, J. K. (1963). *J. Math. Phys.* **4**, 248.

Lebowitz, J. L., Percus, J. K., and Sykes, J. (1969). *Phys. Rev.* **188**, 487.

Leslie, D. C. (1973). *Rep. Progr. Phys.* **36**, 1365.

Levine, H. B., and Birnbaum, G. (1968). *Phys. Rev. Lett.* **20**, 439.

Linnebur, E. J., and Duderstadt, J. J. (1973). *Phys. Fluids* **16**, 665.

Lorentz, H. A. (1916). "The Theory of Electrons," 2nd ed. Teubner, Leipzig (reprinted by Dover, New York, 1952).

Louisell, W. (1964). "Radiation and Noise in Quantum Electronics." McGraw-Hill, New York.

Maeda, H., and Saito, N. (1969). *J. Phys. Soc. Japan* **27**, 984.

Mandel, L. (1958). *Proc. Phys. Soc. (London)* **72**, 1037.

Mandel, L. (1967). *In Symp. Mod. Opt. 1967.* Polytechnic Press, New York.

Mandel, L. (1969). *Phys. Rev.* **181**, 75.

Mandel, L., and Wolf, E. (1965). *Rev. Mod. Phys.* **37**, 231.

Mandel, L., and Wolf, E. (1973). *Opt. Commun.* **8**, 95.

Mandel, L., Sudarshan, E. C. G., and Wolf, E. (1964). *Proc. Phys. Soc. (London)* **84**, 435.

Mandel'shtam, L. I. (1926). *Zh. Russ. Fiz. Khim. Obshchestva* **58**, 381.

Marcuse, D. (1972). "Light Transmission Optics." Van Nostrand-Reinhold, Princeton, New Jersey.

Mazur, P. (1958). *In* "Advances in Chemical Physics" (I. Prigogine and S. A. Rice, eds.), Vol. 1. Wiley, New York.

Mazur, P., and Mandel, M. (1956). *Physica* **22**, 289.

Mazur, P., and Terwiel, R. H. (1964). *Physica* **30**, 625.

Mountain, R. D. (1966). *Rev. Mod. Phys.* **38**, 205.

Oberman, C., Ron, A., and Dawson, J. (1962). *Phys. Fluids* **5**, 1514.

Onsager, L. (1931). *Phys. Rev.* **37**, 405; **38**, 2265.

Ornstein, L. S., and Zernike, F. (1918). *Phys. Z.* **19**, 134.

Oxtoby, D. W., and Gelbart, W. M. (1974a). *J. Chem. Phys.* **60**, 3359.

Oxtoby, D. W., and Gelbart, W. M. (1974b). *Phys. Rev.* **A10**, 738.

Page, C. H. (1952). *J. Appl. Phys.* **23**, 103.

Pattanayak, D. N., and Wolf, E. (1972). *Opt. Commun.* **6**, 217.

Pecora, R. (1964). *J. Chem. Phys.* **40**, 1604.

Peřina, J. (1971). "Coherence of Light." Van Nostrand-Reinhold, Princeton, New Jersey.

Phillips, M. (1962). *In* "Handbuch der Physick," Vol. IV. Springer-Verlag, Berlin.

Pike, E. R. (1974). *In* "Photon Counting and Light Beating Spectroscopy" (H. Z. Cummins and E. R. Pike, eds.). Plenum Press, New York.

Pike, E. R., and Jakeman, E. (1974). *In* "Advances in Quantum Electronics" (D. Goodwin, ed.), Vol. II. Academic Press, New York.

Pike, E. R., Jackson, D. A., Bourke, P. J., and Page, D. I. (1968). *J. Phys. E* (*J. Sci. Instrum.*) **1**, 727.

Purcell, E. M. (1956). *Nature* (*London*) **178**, 1449.

Raman, C. V. (1928). *Indian J. Phys.* **2**, 387.

Rayleigh, Lord (1881). *Phil. Mag.* **12**, 81.

Rice, S. O. (1954). *In* "Selected Papers on Noise and Stochastic Processes" (N. Wax, ed.). Dover, New York.

Rosenfeld, L. (1951). "Theory of Electrons." North-Holland Publ., Amsterdam.

Rytov, S. M. (1957). *Zh. Esperim. Teor. Fiz.* **33**, 514 [*English Transl.: Sov. Phys.-JET P* **6**, 401 (1958)].

Schaefer, D. W. (1973). *Science* **180**, 1293.

Schaefer, D. W., and Berne, B. J. (1972). *Phys. Rev. Lett.* **28**, 475.

Schaefer, D. W., and Pusey, P. N. (1972). *Phys. Rev. Lett.* **29**, 843.

Schaefer, D. W., and Pusey, P. N. (1973a). *J. Phys. A* (*Gen. Phys.*) **6**, 490.

Schaefer, D. W., and Pusey, P. N. (1973b). *In* "Coherence and Quantum Optics" (L. Mandel and E. Wolf, eds.) Plenum Press, New York.

Schaefer, D. W., Benedek, G. B., Schofield, P., and Bradford, E. (1971). *J. Chem. Phys.* **55**, 3884.

Scudieri, F., and Bertolotti, M. (1974). *J. Opt. Soc. Am.* **64**, 776.

Scudieri, F., Bertolotti, M., and Bartolino, R. (1974). *Appl. Opt.* **13**, 181.

Scully, M. (1969). *In Int. School Phys. Enrico Fermi, Varenna, Italy. 1967.* Academic Press, New York.

Shen, Y. R. (1967). *Phys, Rev.* **155**, 921.

Shen, Y. R. (1969). *In Int. School Phys. Enrico Fermi, Varenna, Italy, 1967.* Academic Press, New York.

Smolouchosky, M. (1908). *Ann. Phys.* **25**, 205.

Sommerfeld, A. (1956). "Thermodynamics and Statistical Mechanics." Academic Press, New York.

Steele, W. A., and Pecora, R. (1965). *J. Chem. Phys.* **42**, 1863; **42**, 1872.

Stratton, J. A. (1941). "Electromagnetic Theory." McGraw-Hill, New York.

Sudarshan, E. C. G. (1963). *Phys. Rev. Lett.* **10**, 277.

Swift, J. (1973). *Ann. Phys.* (*N.Y.*) **75**, 1.

Swinney, H. L. (1974). *In* "Photon Counting and Light Beating Spectroscopy" (H. Z. Cummins and E. R. Pike, eds.). Plenum Press, New York.

Tanaka, M. (1968). *Progr. Theor. Phys.* **40**, 975.

Tartaglia, P., and Chen, S. H. (1973). *J. Chem. Phys.* **58**, 4389.

Tyndall, J. (1869a). *Phil. Mag.* **37**, 384.

Tyndall, J. (1869b). *Phil. Mag.* **38**, 156.

van de Hulst, H. C. (1957). "Light Scattering by Small Particles." Wiley, New York.

van Hove, L. (1954). *Phys. Rev.* **95**, 249.

van Kampen, N. G. (1964). *Phys. Rev.* **135**, A362.

van Kampen, N. G. (1969). *In Int. School Phys. Enrico Fermi, Varenna, Italy, 1967.* Academic Press, New York.

Villars, F., and Weisskopf, V. F. (1954). *Phys. Rev.* **94**, 232.

Whitelaw, J. H. (1973). Turbulence Models and Their Experimental Verification. Course given at the Dept. Mech. Eng., Imperial College, London.

Wolf, E. (1963). *In Proc. Symp. Opt. Masers.* Wiley, New York.

Yeh, Y., and Cummins, H. Z. (1964). *Appl. Phys. Lett.* **4**, 176.

Yvon, Y. (1937). "Actualités Scientifiques et Industrielles," Nos. 542, 543. Herman, Paris.

Zimm, B. H. (1945). *J. Chem. Phys.* **13**, 141.

Index